Lecture Notes in Economics and Mathematical Systems

659

For further volumes:
http://www.springer.com/series/300

Lecture Notes in Economics
and Mathematical Systems

530

Founding Editors:

M. Beckmann
H. P. Künzi

Managing Editors:

Prof. Dr. G. Fandel
Fachbereich Wirtschaftswissenschaften
Fernuniversität Hagen
Feithstr. 140/AVZ II, 58084 Hagen, Germany

Prof. Dr. W. Trockel
Institut für Mathematische Wirtschaftsforschung (IMW)
Universität Bielefeld
Universitätsstr. 25, 33615 Bielefeld, Germany

Editorial Board:

A. Basile, A. Drexl, G. Feichtinger, W. Güth, K. Inderfurth, P. Korhonen,
W. Kürsten, U. Schittko

Sarah Bretschneider

Mathematical Models for Evacuation Planning in Urban Areas

 Springer

Dipl. Math. Sarah Bretschneider
Solingen
Germany

ISSN 0075-8442
ISBN 978-3-642-28758-9   ISBN 978-3-642-28759-6 (eBook)
DOI 10.1007/978-3-642-28759-6
Springer Heidelberg New York Dordrecht London

Library of Congress Control Number: 2012943378

Printed on acid-free paper

Springer is part of Springer Science+Business Media (www.springer.com)

Preface

This book results from my work as a research assistant at the university of Duisburg-Essen. The research was partly supported by the WestLB—Stiftung Zukunft NRW.

Evacuation planning involves many tasks, like the determination of the evacuation area, safe zones, and appropriate shelters, as well as the coordination and training of the emergency staff or the planning of the routing and regulation of traffic on the evacuation routes. This research concentrates on the optimization of the routing for the case of an evacuation of urban areas with high density and complex street network structure. Especially, the properties of intersections are investigated and heuristic methods are developed to solve the resulting complex mixed-integer linear optimization programs.

Without the support of my supervisors, colleagues, friends, and my family, it would not have been possible to finish the research and the book in this way. I want to thank my doctoral supervisors Alf Kimms and Rüdiger Schultz for providing me their guidance and advice; special thanks to Alf Kimms who gave me the opportunity to work at the project "Evacuation Problems in Major Cities and Urban Areas". It was a great honor to work with my workmates Demet Çetiner, Michaela Graf, Klaus-Christian Maassen, and Julia Schaaff as well as the newer ones Christoph Reiners and Waldemar Grauberger. It was a great working atmosphere and thank you for going together through the adventures of university, like teaching, handling administrative jobs, or holding talks on conferences.

Especially, I want to express my thanks to my husband Michael for his love and patience. I have so much luck that he is with me. Finally, I thank my family for their support and for sharing all the smaller and greater steps of my research.

Solingen, Germany   Sarah Bretschneider
January 2012

Contents

List of Figures

List of Tables

Chapter 1
Introduction

The population of an urban area may be in danger due to natural or man-made disasters like floods, hurricanes, chemical or nuclear accidents. This requires decisions to protect health and lives of the affected population. A measure to protect the population may be the evacuation of the affected area. Congested urban areas have usually complex street networks that are composed of many intersections with streets which connect them. There are various types of intersections and the streets consist of differing numbers of lanes. The population density of a congested urban area is usually high and the street network is already used to capacity during rush hour traffic. The basic idea is to reorganize the traffic routing of an urban area for the case of an emergency mass evacuation such that aspects of the evacuation like safety, avoidance of delays and/or the total system travel time are taken into account. In this work, the reorganization of the traffic routing will be modeled and solved with tools of mathematical programming.

1.1 Evacuation Planning

A regional evacuation is defined by governments and different organization in similar ways. The United States Department of Homeland Security (DHS, 2004, p. 65) defines an evacuation of a region as

> "Organized, phased, and supervised withdrawal, dispersal, or removal of civilians from dangerous or potentially dangerous areas, and their reception and care in safe areas."

Similar definitions are given for evacuation of an area for example by Müller (1998) (German Bundesamt für Zivilschutz).

Beside the aspects of removal and transport, evacuation planning also needs to incorporate the protection and care of the population that has to leave their residence. We restrict ourselves to the reorganization of traffic routing of an urban street network in case of an emergency mass evacuation.

S. Bretschneider, *Mathematical Models for Evacuation Planning in Urban Areas*,
Lecture Notes in Economics and Mathematical Systems 659,
DOI 10.1007/978-3-642-28759-6_1, © Springer-Verlag Berlin Heidelberg 2013

Evacuation is not always a feasible decision to protect the lives of residents. Another possibility is to order in-place sheltering if an area is affected. FEMA (1996) (United States Federal Emergency Management Agency) considers different examples of hazards where an evacuation may be reasonable and/or necessary. Hurricanes and dam failures are typical examples where an evacuation decision is always an option. In the case of a chemical accidents, the decision about the better alternative between evacuation or in-place sheltering depends, for example, on the type of chemical accident and/or the dispersion of the pollution, see also Müller (1998). In the case of a tornado, evacuation is not a usual measure because the prediction of the track of a tornado is not possible. Sheltering in-place is the recommended option. We refer to FEMA (1996) for more details of emergency planning and examples of hazards.

Criteria for an evacuation of an area are for example the predicted track of a hurricane. Regnier (2008) investigates problems of evacuation decisions concerning whole regions or cities due to hurricanes and therefore she develops models that estimate the uncertainty of forecasts to support the decision maker.

Besides the nature and size of an emergency and the typical criteria of a hazard, a useful and important measure is the estimated evacuation time of the affected area. With the value of the estimated evacuation time and the prediction of when polluted air will strike a certain region, a decision maker has the information to provide support in deciding which measures are appropriate in a certain case: Evacuating or sheltering in-place and preparing the shelter adequately. Sheltering in-place may be a better option if the hazard is going to strike an area while the people are on the streets during an evacuation.

Evacuation time can be defined as the time needed to complete an evacuation process (see e.g. Hamacher and Tjandra, 2001) and includes the components initial warning time, individual's evacuation preparation time, network clearance time and evacuation verification time (Sheffi et al., 1982). We will concentrate on the network clearance time which is understood as the time needed for a given number of evacuees to leave the network of the evacuation. Throughout the text we will call the network clearance time *evacuation time*.

With our results a lower bound of the total evacuation time can be computed and used to decide on an evacuation.

Evacuation is a tool to save lives of people and is incorporated in emergency management. Comprehensive emergency management consists of four phases that may overlap. For more details see for example FEMA (1996), Bumgarner (2008) or Bullock et al. (2008). In the following, the brief description of emergency management functions is based on the "all-Hazard Guide" (FEMA, 1996). Emergency planning for a certain region involves the identification of hazards, their risks, their impacts, consideration of different scenarios, determination of responsibilities, coordination of the different responsibilities and the corresponding coordinators. The emergency management phases or functions are: mitigation, preparedness, response and recovery.

Mitigation deals with the reduction or elimination of the effects of a hazard. This includes for example the adjustment and enforcement of building code requirements in high-hazard areas or the determination of locations of shelters in emergencies within and outside high-hazard areas. Ongoing efforts may be to bring mitigation measures to public attention such that the population that may be affected is aware of what they can do to avoid risks.

The preparedness phase includes the efforts to be prepared for hazards that will probably threaten the area of the corresponding jurisdiction. The authorities and those responsible for emergency duties have to be organized and prepared for when a hazard occurs. Staff have to be recruited and trained. All necessary equipment, resources, facilities, plans and policies have to be determined and prepared. This also includes the planning for an evacuation. The necessary tasks have to be coordinated between the organizations involved, i.e. the government and nongovernmental organizations (like the Red Cross).

The response phase involves actions that are necessary to save lives and property right after the onset of an emergency or disaster: emergency management staff have to be informed, the population has to be warned and, if necessary, evacuated and/or sheltered. Response actions also include preservation of the rule of law, preparation of medical treatment and assessment of the situation and any damage to determine further actions.

Short term efforts of the recovery phase seek to cover basic human and societal needs and to provide necessary lifeline systems like water or power. Long term goals are the rebuilding of the infrastructure while incorporating mitigation efforts and restoring the social and economic life to normal.

Emergency evacuation planning involves the planning of the required supplies, equipment and personnel such that the planning and coordination of transportation and accommodation of the evacuees can be accomplished. That means planning of mass evacuation includes the determination of the evacuation area, safe zones and shelters. Transportation planning is necessary to ensure that all evacuees have the opportunity to leave the evacuation zone. This includes the planning of the routing and regulation of traffic on evacuation routes as well as providing adequate transport resources for people that do not have the possibility to self-evacuate, like residents of hospitals or prisons. In shelters, adequate resources like food, water, beds, medical and sanitary supplies and medical support have to be provided. Registration processes for finding relatives, the reunification of families and safety have to be established. To encourage as many people as possible to leave the evacuation zone, household pets and service animals are evacuated to appropriate shelters. Both the conditions of return and the actual return of the evacuees to the affected area have to be determined, planned and coordinated. For more details of evacuation planning see e.g. DHS (2008) and Reichert (2002).

We concentrate on the determination of the traffic routing of a mass evacuation in the planning phase. The traffic routing underlies certain rules that support a more safe evacuation process under the objectives of minimizing the average evacuation time per vehicle or minimization of the evacuation time. The locations of the shelters outside the evacuation zone are input data for the optimization models which will be presented in the following chapters as are the number of usable lanes, the number of evacuees and capacity values. Different scenarios need to be considered, i.e. instances with different exit-patterns, capacities or evacuation areas. The presented models are to be used before a hazard or disaster threatens an area in such a way that an evacuation is necessary. The results like the traffic routing or the resulting estimated evacuation time may be used for decision support. Information about the changed traffic routing in case of an evacuation can be plotted on route maps and may be published in information brochures, evacuation handbooks or on the web. This is especially important for high-hazard areas like coastal areas with hurricane season. It has to be ensured that the public stays informed and efforts are made to explain the contraflow plan (see e.g. Wolshon, 2001).

1.2   Basic Network Flow Problems

The traffic flow of evacuees is modeled in this work with linear models that are based on the linear dynamic network flow model presented by Ford and Fulkerson (1958). This approach incorporates time explicitly. The time horizon is discretized into time periods of equal length and the flow needs a certain number of periods to traverse the arcs in the network.

In the following a in two variants will be presented which builds the basis of the evacuation models. The flow propagation is modeled with flow conservation, capacity and flow enforcement constraints. Initial flow located in source nodes, have to reach sink nodes under predefined constraints and a predefined objective. The generic part of a dynamic network flow model consists of the flow conservation constraints, a capacity constraint regarding the inflow capacity and flow enforcement aspects. The inflow may have different values on an arc in different points in time and the flow on arcs needs a certain amount of time to travel through an arc. The basic objectives that are interesting for evacuation planning are the minimization of the evacuation time (network clearing time), the minimization of the average evacuation time of every vehicle or a more general objective the minimization of the weighted sum of flows entering the arcs leading to the super sink where the weights are increasing in time (i.e. the later flow arrives at a sink the higher the penalty) and the maximization of the total flow entering the sinks during a given time horizon with the earliest arrival property. The earliest arrival property states that for every considered point in time t the cumulative amount of flow entering the sink in the points in time $t' \leq t$ is maximized. We concentrate on minimizing the average evacuation time of every vehicle loaded with a certain amount of evacuees and present some results on minimizing the evacuation time.

Let $\mathbf{G} = (\mathcal{N}, \mathcal{A})$ be a static directed network defined as a set of nodes $\mathcal{N}$ and a set of arcs $\mathcal{A} \subseteq \mathcal{N} \times \mathcal{N}$ including ordered pairs of nodes. The static network $\mathbf{G}$ represents the considered street network with the artificial super nodes and the corresponding arcs.

The set of nodes $\mathcal{N}$ includes a set of source nodes $\mathcal{S} \subset \mathcal{N}$ and a set of sink nodes $\mathcal{D} \subset \mathcal{N}$. Each source $s \in \mathcal{S}$ is assigned a certain amount of supplies $O(s)$, i.e. the number of vehicles initially in the source s, $s \in \mathcal{S}$. The cumulated demand of the sinks is equal to the sum of all the supplies of the sources, i.e. there is no assignment of evacuees to a sink in advance. Also super nodes, the super source S and the super sink D, are within the set of nodes $\mathcal{N}$, i.e. $\mathcal{N}$ that incorporate the street network nodes as well as the super source S and the super sink D. The street network nodes may represent intersections or certain significant points in the street network like the starting points of evacuees, the sources, or entrances of a safe zone, or the sinks.

The set of arcs $\mathcal{A}$ includes the arcs emerging from the super source S to the sources, i.e. $\mathcal{A}_S := \{(S, s) : s \in \mathcal{S}\}$, and the arcs emerging from sinks to the super sink D, i.e. $\mathcal{A}_D := \{(d, D) : d \in \mathcal{D}\}$. That means, the set of arcs $\mathcal{A}$ includes the street network arcs and the artificial arcs connecting the sources and sinks with the corresponding super node. The attributes defined on the arcs are the travel time $\tau(i, j)$ that is needed to traverse the arc $(i, j) \in \mathcal{A}$ and the inflow capacity $c^{in}(i, j)$ that limit the number of flow units which can enter arc $(i, j) \in \mathcal{A}$ in every point in time. The travel times of arcs emerging from the super source S and pointing to the super sink D are in general set to 0 and the inflow capacity is infinite, i.e. there are no capacity restrictions on these arcs. The time horizon is divided into $T \in \mathbb{N}$ time periods of equal length, i.e. $T + 1$ points in time are considered. Let $\mathcal{T} = \{0, 1, \cdots, T\}$ be the set of considered points in time.

The time-expanded network $\mathbf{G}^T$ is constructed based on the static network $\mathbf{G}$, the set $\mathcal{T}$ and the travel times. The travel times are considered as constant. The generation of the time-expanded network with time-dependent travel time is analog to the generation of the time-expanded network with constant travel time. The time-expanded network $\mathbf{G}^T = (\mathcal{N}^T, \mathcal{A}^T)$ is built as described in the following (compare e.g. Ford and Fulkerson, 1958; Hamacher and Tjandra, 2001). The set $\mathcal{N}^T$ of nodes is defined as

$$\mathcal{N}^T := \{i(t) : i \in \mathcal{N}, t \in \mathcal{T}\}.$$

The set of arcs $\mathcal{A}^T$ is divided into movement arcs $\mathcal{A}^M$ and holdover arcs $\mathcal{A}^H$:

$$\mathcal{A}^M := \{(i(t), j(t')) : (i, j) \in \mathcal{A}, t \in \mathcal{T}, t' = t + \tau(i, j, t) \in \mathcal{T}\}$$
$$\mathcal{A}^H := \{(i(t), i(t')) : i \in \mathcal{N}\backslash\{S, D\}, t \in \mathcal{T}, t' = t + 1 \in \mathcal{T}\}$$

Holdover arcs represent the waiting period in nodes. The time needed to traverse a holdover arc is generally set to one and a capacity on these arcs may be considered. The inflow capacity c^{in} assigned to the arcs of the static network are also associated with the corresponding movement arcs of the time-expanded network. The inflow capacity may be considered as time-varying.

Figure 1.1 gives an example of a static network and the corresponding time-expanded network with $T = 6$. The values next to the arcs of the static network in Fig. 1.1a depict the travel times on the corresponding arcs. The horizontal dashed arcs in the time-expanded network (see Fig. 1.1b) represent the holdover arcs and the solid arcs represent the movement arcs. I.e. in this approach a snapshot of the street network is considered for every point in time.

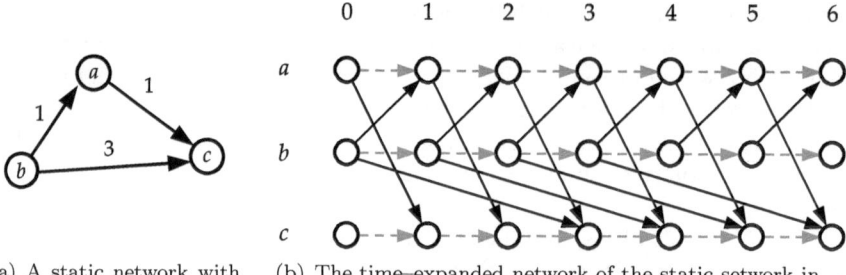

(a) A static network with   (b) The time–expanded network of the static setwork in
travel times on arcs   Figure 1.1(a) with $T = 6$

Fig. 1.1 Illustration of a static and the corresponding time-expanded network

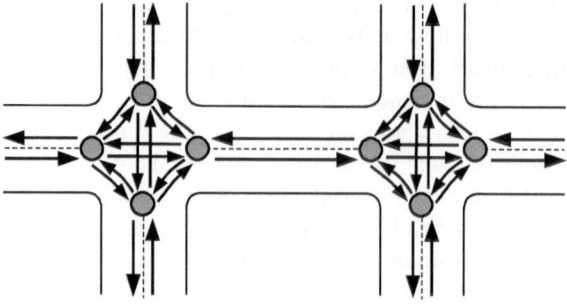

Fig. 1.2 An illustration of a graph of a street network

The developed evacuation models that will be presented in the following chapters are based on a discrete time dynamic network flow model, where the time-expanded network is incorporated. In this work two alternatives of flow models with constant attributes are considered. To build a basic simple model the waiting time in nodes is not considered in the first flow model. In the second flow model, an extension of the first one, "waiting on arcs" is included. In other words, the waiting of flow depends on the direction the flow came from. The first model will be introduced in Sect. 1.2.1, the second one in Sect. 1.2.2. "Waiting on arcs" seems more reasonable for the representation traffic flow where different directions, i.e. the arcs, have waiting capacities than the waiting in nodes as in the "classical" dynamic network flow model like the models in Ford and Fulkerson (1958), Chalmet et al. (1982), Hoppe and Tardos (1994) or Hamacher and Tjandra (2001). The last three especially want to map building evacuation. In the case of building evacuation, nodes represent rooms, workplaces or stairwells, where a certain amount of waiting space for people exists and it is reasonable to model a scarce capacity.

For urban evacuation, we choose to generate the graph of street networks as follows: Arcs depict a certain direction of a street section and nodes specific points like points where two or more different street section adjoin each other. The nodes have no real capacity, they just mark points that have to be passed. Figure 1.2 shows

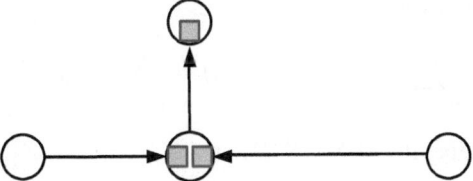

Fig. 1.3 An illustration of "waiting on arcs" described as waiting in boxes in nodes

a small street network with two four-leg intersections. The arcs represent a direction, the nodes points to pass, to enter or to exit an intersection.

The considered "waiting on arcs" models the waiting at the end of an arc. In other words, if flow enters the head node after the travel time τ, then there is the possibility to wait there in boxes that are only reserved for flow coming from the corresponding arc. For an illustration, see Fig. 1.3. The distinction between these flows is important in mapping waiting along arcs, similarly to a street section. What is realized is "waiting on an arc" with another decision variable that is defined on the corresponding arc at a certain point in time.

1.2.1 A Generic Network Flow Model

The following model is a generic formulation of the minimum cost dynamic network flow problem that builds the basis of the developed evacuation models (similar to the minimum cost flow model presented in Hamacher and Tjandra 2001). The possibility of waiting in nodes is only given in the super source S. The number of vehicles allocated at the beginning of the considered time horizon in the sources $s \in S$ is donated by $O(s)$. Decision variables are the inflow $x(i, j, t)$ on arc $(i, j) \in \mathcal{A}$ per point in time $t \in T$ (i.e. inflow rate) that leave node i in time t and enters node j in point in time $t + \tau(i, j)$. For the ease of notation $x(i, j, t)$ is set to 0 for $t < 0$ and $t + \tau(i, j) > T$, for arcs (i, j) in $\mathcal{A}$. Let γ represent real-valued weights associated with arcs and points in time, i.e. $\gamma : \mathcal{A} \times T \to \mathbb{R}$. The generic network flow model without waiting is described in the following.
Objective function:

$$\min \sum_{(i,j) \in \mathcal{A}} \sum_{t \in T} \gamma(i, j, t) \cdot x(i, j, t) \tag{1.1}$$

Flow enforcement constraints:

$$\sum_{t \in T} x(S, s, t) = O(s) \quad \text{for all } s \in S \tag{1.2}$$

$$\sum_{t \in T} \sum_{d \in D} x(d, D, t) = \sum_{s \in S} O(s) \tag{1.3}$$

Flow conservation constraints:

$$\sum_{(i,j)\in\mathcal{A}} x(i,j,t-\tau(i,j)) = \sum_{(j,i)\in\mathcal{A}} x(j,i,t) \tag{1.4}$$

for all $j \in \mathcal{N}\backslash\{S,D\}$; $t \in \mathcal{T}$

Capacity constraints:

$$0 \le x(i,j,t) \le c^{in}(i,j,t) \text{ for all } (i,j) \in \mathcal{A}; t \in \mathcal{T} \tag{1.5}$$

The objective function (1.1) is the weighted sum of flows with weights depending on arcs and points in time. All vehicles have to leave the corresponding source within the time horizon (1.2) and all vehicles have to reach a safe zone within the time horizon (1.3). The flow conservation constraint (1.4) states: the total flow that enters a node (without the super nodes) at point in time t has to leave this node at point in time t. Constraint (1.5) restricts the inflow per point in time on every arc. Note, with the restriction of the inflow, the total flow on arc $(i,j) \in \mathcal{A}$ is restricted by $\tau(i,j) \cdot c^{in}(i,j)$.

Incorporating Waiting in Nodes

To include waiting in nodes, waiting variables $y(i,t) \in \mathbb{R}_0^+$, $i \in \mathcal{N}$, $t \in \mathcal{T}$ and a capacity constraint on the waiting variables per time may be added as well as the flow conservation constraints have to be changed the following way.
Flow conservation constraint with waiting in nodes:

$$\sum_{(i,j)\in\mathcal{A}} x(i,j,t-\tau(i,j)) + y(j,t-1) = \sum_{(j,i)\in\mathcal{A}} x(j,i,t) + y(i,t) \tag{1.6}$$

for all $j \in \mathcal{N}\backslash\{S,D\}$; $t \in \mathcal{T}$

Capacity limit on the number of vehicles in nodes per point in time:

$$0 \le y(i,t) \le c^w(i,t) \quad \text{for all } i \in \mathcal{N}\backslash\{S,D\}; t \in \mathcal{T} : t+1 \le T \tag{1.7}$$

where $c^w(i,t)$ denotes the waiting capacity in nodes $i \in \mathcal{N}\backslash\{S,D\}$ per $t \in \mathcal{T}$.

1.2.2   A Generic Network Flow Model with Waiting on Arcs

Another possible way to model waiting for traffic flow is to allow "waiting on arcs" as described above. I.e. the direction from which the flow arrives is significant. Beside the flow conservation constraints and capacity constraints, the constraints that allow the flow, which has traveled on an arc, to wait on this arc have to be included. Then the travel time can be interpreted as the minimum travel time on a street segment. The decision variable $w(i, j, t) \in \mathbb{R}_0^+$ stands for the number of vehicles that begin to wait on arc $(i, j) \in \mathcal{A}$ at point in time t and end waiting at point in time $t + 1$. After this time unit, the flow can wait for another time unit or leave node j through an arc $(j, k) \in \mathcal{A}$. For the ease of notation, we set $w(i, j, t) :=$ 0 for $t < 0$. And let $w(i, j, t) := 0$ for $(i, j) \in \mathcal{A}_S \cup \mathcal{A}_D \cup \{(i, d) \in \mathcal{A} : d \in \mathcal{D}\}$. The generic network flow model with waiting on arcs is described as follows.

Objective function:

$$\min \sum_{(i,j) \in \mathcal{A}} \sum_{t \in \mathcal{T}} \gamma(i, j, t) \cdot x(i, j, t) \qquad (1.8)$$

Flow enforcement constraints:

$$\sum_{t \in \mathcal{T}} x(S, s, t) = O(s) \quad \text{for all } s \in \mathcal{S} \qquad (1.9)$$

$$\sum_{t \in \mathcal{T}} \sum_{d \in \mathcal{D}} x(d, D, t) = \sum_{s \in \mathcal{S}} O(s) \qquad (1.10)$$

Flow conservation constraints with the possibility of waiting on arcs:

$$\sum_{(i,j) \in \mathcal{A}} (x(i, j, t - \tau(i, j)) + w(i, j, t - 1)) \qquad (1.11)$$

$$= \sum_{(j,i) \in \mathcal{A}} x(j, i, t) + \sum_{(i,j) \in \mathcal{A}} w(i, j, t) \quad \text{for all } j \in \mathcal{N} \backslash \{S, D\}; t \in \mathcal{T}$$

Capacity constraints:

$$0 \le x(i, j, t) \le c^{in}(i, j, t) \quad \text{for all } (i, j) \in \mathcal{A}; t \in \mathcal{T} : t + \tau(i, j, t) \le T \quad (1.12)$$

$$0 \le w(i, j, t) \le c^w(i, j, t) \quad \text{for all } (i, j) \in \mathcal{A}; t \in \mathcal{T} : t + \tau(i, j, t) \le T \quad (1.13)$$

$$w(i, j, t) \le w(i, j, t - 1) + x(i, j, t - \tau(i, j)) \quad \text{for all } (i, j) \in \mathcal{A}; t \in \mathcal{T} \qquad (1.14)$$

The objective function (1.8), the flow enforcement constraints (1.9) and (1.10) as well as the inflow capacity constraints (1.12) are the same as the corresponding constraints in the generic model without waiting. The flow conservation constraints (1.11) include the possibility of waiting on arcs after a minimum travel time τ. The flow entering node j at time t or waiting in the corresponding "box" in the node until point in time t has to leave the node j in point in time t or enters the corresponding "box" in node j in time t. The constraints (1.14) guarantee that the flow waits just on an arc coming from or waiting already on that arc. The number of flow units waiting on an arc is restricted (1.13), where $c^w(i, j)$ denotes the waiting capacity on arc $(i, j) \in \mathcal{A}$ per time unit.

Incorporating the possibility of waiting is especially useful if the traffic load entering the street network is predetermined and is not part of a decision as in the presented models. An interpretation of waiting on arcs may be congestion.

1.2.3   Objectives

The objective functions considered in this work are valid for both variants of the dynamic network flow model. The objective functions (1.1) and (1.8) presented for the flow models above are the same. We will only refer to the objective function (1.1) of the dynamic flow model without waiting.

Minimization of the Weighted Sum

We will consider special cases of the objective function weighted sum of flows (1.1).

The weighted sum of flows entering a safe zone, where the weights are increasing in time, can be realized with (1.1) as follows: Let the weights be $\gamma(d, D, t) := \gamma(t)$ increasing with points in time $t \in \mathcal{T}$ for arcs leading from sinks to the super sink, i.e. $(d, D) \in \mathcal{A}_D$ and $\gamma(i, j, t) = 0$ for arcs $(i, j) \in \mathcal{A} \backslash \mathcal{A}_D$ and $t \in \mathcal{T}$. I.e. the later the flow arrives at a sink the higher is the penalty $\gamma(t)$. The objective reads with the above weights as follows:

$$\sum_{d \in \mathcal{D}} \sum_{t \in \mathcal{T}} \gamma(t) \cdot x(d, D, t) \qquad (1.15)$$

The average evacuation time per vehicle can be realized with objective (1.1) as follows (compare Hamacher and Tjandra, 2001): Let the weights be $\gamma(d, D, t) = t$ for $(d, D) \in \mathcal{A}_D, t \in \mathcal{T}$ and $\gamma(i, j, t) = 0$ for $(i, j) \in \mathcal{A} \backslash \mathcal{A}_D, t \in \mathcal{T}$, i.e.

$$\sum_{d \in \mathcal{D}} \sum_{t \in \mathcal{T}} t \cdot x(d, D, t) \qquad (1.16)$$

Objective (1.16) describes the total sum of time units the vehicles of the evacuees need to leave the evacuation area. At point in time t, the amount of flow $x(d, D, t)$ has entered safe zone $d \in \mathcal{D}$. The following formula describes the average evacuation time per vehicle, where we assume that the evacuation time starts to count with $t = 0$ for every vehicle:

$$\frac{\sum_{d \in \mathcal{D}} \sum_{t \in \mathcal{T}} t \cdot x(d, D, t)}{\sum_{s \in \mathcal{S}} O(s)}$$

The total number of vehicles $\sum_{s \in \mathcal{S}} O(s)$ is constant. Therefore, just the numerator needs to be considered.

The cost $\gamma(d, D, t) = t$ associated with the arc (d, D), $d \in \mathcal{D}$, and point in time $t \in \mathcal{T}$ is called turnstile cost (see e.g. Chalmet et al. 1982, or Hamacher and Tufekci 1987).

Minimization of the Evacuation Time

To minimize the evacuation time an additional node, the super–super sink D^* and an additional arc, the arc (D, D^*), pointing from the super sink to the super-super sink are added. The minimization of the evacuation time, i.e. the network clearance time, can be realized as follows:

$$\min \quad \Theta \tag{1.17}$$

$$s.t. \quad \Theta \geq t \cdot y_E(t) \quad \text{for all } t \in \mathcal{T} \tag{1.18}$$

$$x(D, D^*, t) \leq c^{in}(D, D^*) y_E(t) \quad \text{for all } t \in \mathcal{T} \tag{1.19}$$

where the decision variables $\Theta \in \mathbb{R}_0^+$ and $y_E(t) \in \{0, 1\}$, $t \in \mathcal{T}$ and a necessary large value $c^{in}(D, D^*) \geq 0$ are associated with arc (D, D^*). The flow conservation constraints have to be adjusted: the considered nodes have to be in the set $\mathcal{N} \backslash \{S, D^*\}$. The problem of the minimization of the clearance time with multiple sources and multiple sinks is called the quickest transshipment problem.

The "price" of finding the minimum evacuation time is that there are $T + 1$ additional binary variables that indicate if flow uses the arc (D, D^*) in a time period $t \in \mathcal{T}$ or not.

Maximization of the Flows

Dynamic maximum flow problems seek to maximize the amount of flow traveling from sources to sinks in a given time horizon. Ford and Fulkerson (1958) considered the single source and single sink case, where no certain supply needs to be assigned because the underlying question is to find the maximum sum of flows that can enter the sink emerging from the source within a given time horizon. Considering this question for the multiple source case, it is possible that just one source sends out the total flow to a sink and all other sources remain unconsidered. An extension of the problem of dynamic maximum flow is to assign supplies and demands to sources and sinks, respectively, that have to be satisfied within a given time horizon and the flow has to fulfill the earliest arrival property (see e.g. Baumann and Skutella 2006). If the time horizon is chosen too small then the problem is infeasible.

We will suggest an alternative problem that is feasible for every given time horizon (if the underlying network is connected and from every source a path to a sink exists) but that takes also the supplies of sources into account. The problem is to find the maximal amount of flow that enters the sinks within a given time horizon when every source can only send up to the assigned number of evacuees/vehicles. The flow has to fulfill the flow conservation and capacity constraints.

The maximization of the flow entering the sinks during a given time horizon can be modeled as follows:

$$\max \qquad \sum_{d \in \mathcal{D}} \sum_{t \in \mathcal{T}} \gamma_{max}(t) \cdot x(d, D, t) \qquad (1.20)$$

where with weights $\gamma_{max}(t)$ decrease with points in time, i.e. flow entering the sinks earlier are rewarded more in sense of a higher weight.

In both presented dynamic network flow models, constraint (1.3) (or (1.10) in the model with waiting on arcs) that states that all vehicles have to reach a destination within the considered time horizon is skipped. If the chosen time horizon is too small, such that not every vehicle can reach a destination, then the instance of the generic dynamic network flow model is infeasible. To achieve a feasible solution for every chosen time horizon with the drawback that probably not all vehicles have left the evacuation zone the map γ can be chosen as $\gamma_{max}(t) = (T + 1) - t$, the constraints (1.3) can be disregarded and constraints (1.2)/(1.9) can be relaxed to $\sum_{t \in \mathcal{T}} x(S, s, t) \leq O(s)$ for all $s \in \mathcal{S}$. A solution of an instance with a smaller time horizon is probably computed faster.

An equivalent formulation can be realized with the objective function (1.1) with the weights $\gamma(d, D, t) := \gamma_{min}(t)$ for $d \in \mathcal{D}$, $t \in \mathcal{T}$ and $\gamma(i, j, t) := 0$ otherwise, where γ_{min} is increasing with points in time and $\gamma_{min}(t) < 0$ for all $t \in \mathcal{T}$:

$$\min \qquad \sum_{d \in \mathcal{D}} \sum_{t \in \mathcal{T}} \gamma_{min}(t) \cdot x(d, D, t) \qquad (1.21)$$

For example, γ could be set to $\gamma_{min}(t) = -T - 1 + t$.

In this work, we will concentrate on the minimization cost flow problem and we will give some results for the problem of minimizing the evacuation time.

1.3 Crossing and Merging Conflicts

The traffic routing of an urban area shall be adjusted for the case of an evacuation. Therefore the available number of lanes are reallocated to the two possible directions of each street segment subject to restrictions that avoid certain traffic conflicts. Traffic management strategies like lane-reversal operations or the use of shoulder lanes of freeways are incorporated in traffic planning for regional evacuation (see e.g. Hobeika and Kim, 1998; Urbina and Wolshon, 2003; DHS, 2008; Peeta and Kalafatas, 2008; Wolshon and McArdle, 2009) to improve the traffic routing in case of an regional mass evacuation. Lane-reversal operations are commonly operated for interstate freeway traffic. We will include the readjustment of urban

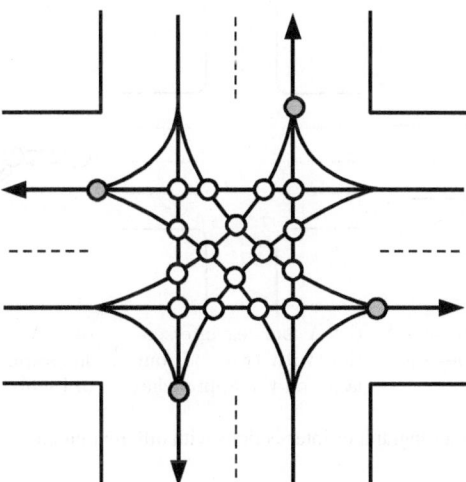

Fig. 1.4 Crossing and merging conflicts within an intersection with one lane on each direction

roadways. Contraflow operations are used for the arterial urban roadways during rush hour periods or special events and do not decrease traffic safety if they are well controlled (Wolshon 2001). As much as possible of the capacity of the usable street network should be used for the evacuation process. Delays and potential accident situations should be reduced. Traffic delays occur most within intersections (see e.g. Cova and Johnson 2002, or Southworth 1991). The right of way or traffic signals have to be obeyed or turning vehicles in front may cause the deceleration of the following vehicles. Traffic conflicts are defined in Rao and Rengaraju (1997, p. 81) as "interactions between vehicles that occur when one or more vehicles take evasive actions, such as braking or weaving, to avoid a collision". The number of intersecting traffic streams affect the number of traffic conflicts within uncontrolled intersections which "can be used as an indicator for possible accidents" (Rao and Rengaraju, 1997, p. 81).

In the following, crossing conflicts and merging conflicts are defined by means of lanes (see e.g. Cova and Johnson 2002): Let a crossing conflict within an intersection be the crossing of lanes and let a merging conflict be the merging of two or more lanes into one lane.

Figure 1.4 illustrates conflicts of a four-leg intersection with one lane on each possible direction. A line depicts a lane. The filled points represent the merging conflicts and the circles represent crossing conflicts.

To avoid delays and reduction of the potential of accidents at and within intersections, crossing conflicts are prohibited and merging conflicts restricted in the traffic routing reassigned for the case of a mass evacuation. The considered street network will be modeled through a direction-based graph. The graph representing the street network consists of arcs that depict directions. For an illustration see Fig. 1.5. An intersection with two available lanes on each street segment is depicted in Fig. 1.5a and the intersection depicted in Fig. 1.5b with a mixed number of lanes.

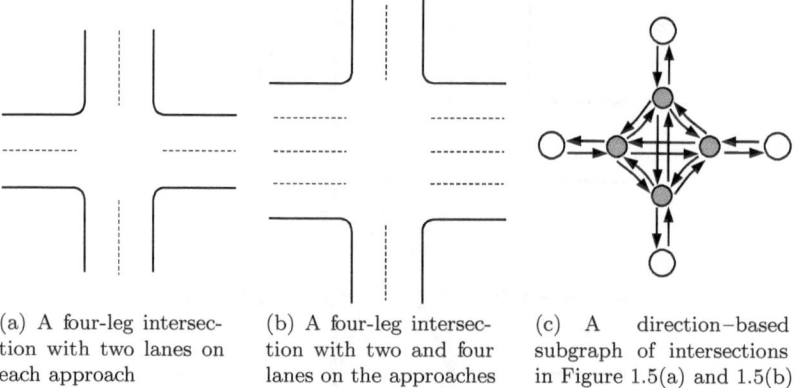

(a) A four-leg intersec-
tion with two lanes on
each approach

(b) A four-leg intersec-
tion with two and four
lanes on the approaches

(c)   A   direction–based
subgraph of intersections
in Figure 1.5(a) and 1.5(b)

Fig. 1.5 Illustration of a subgraph of intersections with different numbers of lanes

The number of legs of an intersection are crucial to generate a subgraph of an intersection in a direction-based sense. Figure 1.5c depicts the subgraph of both intersections.

The crossing conflicts that we consider are crossing conflicts which are lane-based, but which are not directly modeled with the lanes of street segments, they are modeled "direction-basis". The amount of lanes subject to certain constraints (see Chap. 3) can be arranged in such a way that no crossing conflict of lanes occurs. To guarantee a traffic routing without crossing conflicts, the following assumption is necessary: vehicles have to order in the appropriate lanes that correspond to their subsequent turn before they enter the intersection.

1.4   Chapter Synopsis

First, we will consider the single commodity case, i.e. every evacuee has the possibility to self-evacuate, and concentrate on modeling the flow of vehicles from the evacuation area to safe zones. Then, we will include the scenario where not every person has the possibility to leave the evacuation zone and incorporate pick up points where evacuees can enter buses. In the multicommodity case, the flow of vehicles of evacuees and the flow of buses will be modeled.

The text is organized as follows: in Chap. 2 a literature review about evacuation research will be given. It will show that most of the literature concentrates on the optimization and simulation of flows, where the traffic routing, i.e. the capacities, are given, and only a few on the optimization of flow and the traffic routing subject to certain traffic management strategies. In Chap. 3 a mixed-integer urban evacuation model will be presented that restructures the traffic routing. A relaxation-based heuristic approach is presented in Chap. 4 where some traffic routing constraints are relaxed and later rebuilt. Chapter 5 provides a pattern-based model of the urban evacuation problem. The construction of patterns and some properties are presented. The pattern-based evacuation model is solved with a two-staged heuristic approach that considers the network in different aggregation levels (Chap. 6). Chapter 7 introduces a multicommodity urban evacuation model that takes the flow of self evacuating vehicles and the flow of buses into account. Buses can pick up evacuees in predetermined locations to transport them to safe zones. A bus can enter the evacuation zone multiple times if necessary. The multicommodity evacuation problem can be solved with a heuristic that is introduced in Chap. 8. This heuristic operates again on different aggregation levels of the street network. Chapter 9 will conclude this work.

References

Baumann N, Skutella M (2006) Solving evacuation problems efficiently–earliest arrival flows with multiple sources. In: 47th annual IEEE symposium on foundations of computer science (FOCS'06), p 399–410

Bullock JA, Haddow G, Coppola DP, Yeletaysi S (2008) Introduction to Homeland security: principles of all-hazards response, 3rd edn. Butterworth-Heinemann, Oxford

Bumgarner JB (2008) Emergency management: a reference handbook. ABC-CLIO, Santa Barbara, CA

Chalmet LG, Francis RL, Saunders PB (1982) Network models for building evacuation. Manag Sci 28(1):86–105

Cova TJ, Johnson JP (2002) Microsimulation of neighborhood evacuations in the urban-wildland interface. Environ Plann A 34:2211–2229

DHS (2004) National response plan. U.S. Department of Homeland Security, USA

DHS (2008) Mass evacuation incident annex, U.S. Department of Homeland Security. http://www.fema.gov/pdf/emergency/nrf/nrf_massevacuationincidentannex.pdf. Accessed 19 Oct 2010

FEMA (1996) Guide for all-hazard emergency operations planning, Federal Emergency Management Agency, USA

Ford LR Jr, Fulkerson DR (1958) Constructing maximal dynamic flows from static flows. Oper Res 6(3):419–433

Hamacher H, Tjandra S (2001) Mathematical modeling of evacuation problems: a state of art. Berichte des Frauenhofer ITWM, Nr. 24. http://www.itwm.fraunhofer.de/fileadmin/ITWM-Media/Zentral/Pdf/Berichte_ITWM/2001/bericht24.pdf. Accessed 18 Oct 2006

Hamacher HW, Tufekci S (1987) On the use of lexicographic min cost flows in evacuation modeling. Nav Res Logist 34:487–503

Hobeika A, Kim C (1998) Comparison of traffic assignments in evacuation modeling. IEEE Trans Eng Manag 45(72):192–198

Hoppe B, Tardos E (1994) Polynomial time algorithms for some evacuation problems. In: Proceedings of the fifth annual ACM-SIAM symposium on discrete algorithms 1994, p 433 441, Arlington, VA, 1994

Müller G (1998) Kritierien für Evakuierungsempfehlungen bei Chemikalienfreisetzungen, Bundesamt für Zivilschutz, Zivilschutz-Forschung, Band 32, Bonn

Peeta S, Kalafatas G (2008) Primary emergency routes for transportation security. Joint transportation research program, Paper 323. http://docs.lib.purdue.edu/jtrp/323

Rao VT, Rengaraju VR (1997) Probabilistic model for conflicts at urban uncontrolled intersection. J Transport Eng 123(1):81–84

Regnier E (2008) Public evacuation decisions and hurricane track uncertainty. Manag Sci 54(1):16–28

Reichert D (2002) Evakuierung und Unterbringung, Rescue 2002, Fachkongress für interdisziplinäre Zusammenarbeit im Rettungwesen und in der Gefahrenabwehr. http://www.landkreis-dillingen.de/lra/kats/evakuierung.pdf. Accessed 18 Oct 2010

Sheffi Y, Mahmassani H, Powell WB (1982) A transportation network evacuation model. Transport
    Res A 16(3):209–218
Southworth F (1991) Regional evacuation modeling: a state-of-the-art review, Oak Ridge National
    Laboratory ORNL-11740, Tennessee, 1991
Urbina E, Wolshon B (2003) National review of hurricane evacuation plans and policies: a
    comparison and contrast of state practices. Transport Res A Pol Pract 37:257–275
Wolshon B (2001) "One-way-out": contraflow freeway operation for hurricane evacuation. Nat
    Hazards Rev 2(3):105–112
Wolshon B, McArdle B (2009) Temporospatial analysis of hurricane katrina regional evacuation
    traffic patterns. J Infrastruct Syst 15(1):12–20

Chapter 2
Literature Review

Evacuation plans are developed for buildings, ships, stadiums or districts, cities or whole sub-national region. In the following literature review, mathematical programming and simulation approaches that consider, mainly, the evacuation planning concerning the evacuation from regions like districts, towns or regions will be presented. Recently, there have been a number of articles concerned with evacuation planning and evacuation support. They could be divided into flow-based optimization approaches which seek to compute an optimal solution of certain objectives and simulation approaches that evaluate an existing evacuation plan.

We will concentrate on dynamic network flow models. Traffic assignment models will be briefly introduced and presented when introducing approaches that apply traffic assignment models to evacuation planning.

2.1 Flow-Based Optimization Models

Dynamic network flow problems and traffic assignment problems are applicable to model flow for evacuation planning. In the following, general approaches of dynamic network flows will be presented and specific approaches of dynamic network flow models and traffic assignment models which are constructed for or applied to evacuation planning will be summarized.

Dynamic network flow models can be used to describe evacuation problems. The question of how to obtain the quickest flow with multiple sources is often referred to as the evacuation problem (see e.g. Hoppe and Tardos, 1994). Dynamic network flow problems are generally applied to model network flow problems where the amount of flow on an arc may vary over time and flow needs a certain amount of time to traverse an arc. In the case of a discrete time model, the considered time horizon is divided into time periods of equal length. Every node and every arc is considered in every point in time. The term *dynamic*, in the case of dynamic network flows, refers to the characteristic of the inflow on an arc which may take different

S. Bretschneider, *Mathematical Models for Evacuation Planning in Urban Areas*,   17
Lecture Notes in Economics and Mathematical Systems 659,
DOI 10.1007/978-3-642-28759-6_2, © Springer-Verlag Berlin Heidelberg 2013

values in different points in time. The parameters are known in advance and no new information about the values of parameters are updated during solution. Dynamic network flows are also called flows over time.

A dynamic network is defined as a directed graph $G = (N,A)$ with non-negative capacities and non-negative, integer transit times on arcs. Two subsets of nodes, the set of sources and the set of sinks, are defined. A dynamic feasible flow satisfies flow conservation constraints and is called feasible if the inflow capacity is regarded. The flow or inflow rate on an arc is the amount of flow in arc (i, j) that leaves node i at point in time t, arrives j at point in time t plus the corresponding travel time. Typical dynamic network flow problems are to find a feasible flow traveling from sources to sinks under certain supply/demand constraints and fulfilling flow conservation constraints while optimizing an objective.

A variety of dynamic network flow problems are considered for evacuation planning: the dynamic maximum flow problem, the earliest arrival flow problem (also known as universal maximum flow problem), the quickest flow or the transshipment problem, or a dynamic minimum cost flow problem. Variations and extensions of the dynamic flow problems correspond to the number of sources and sinks, to the dependencies of attributes of the arcs and nodes, e.g. constant, time-dependent or flow-dependent capacities or transit times on arcs, or as well as additional constraints and/or variables. Dynamic network flow problems model time in discrete time steps or continuously.

The traffic assignment problem seeks to determine flows on links of a given road network under certain optimality or equilibrium conditions, where a flow rate for origin-destination pairs, i.e. OD-pairs, of nodes for every considered point in time is given. The departure flow rate for an OD-pair at time t determines the amount of flow leaving the origin node to travel to the destination node at time t. When the given flow rate for the OD-pairs is constant over a long horizon, the traffic assignment problem is called static, otherwise dynamic (Merchant and Nemhauser, 1978). Models that describe the dynamic traffic assignment problem generally claim to model traffic phenomena like congestion or shockwaves realistically or at least approximate them. Traffic assignment problems are used in simulation approaches where beside the solution of a traffic assignment problem, steps like the determination of the OD-matrix or appropriate path generations have to be implemented. Dynamic traffic assignment problems are used to study evacuation problems as part of a simulation approach (see Sect. 2.4) and as linear optimization problems that are often based on the cell transmission model developed by Daganzo (1994) and Daganzo (1995).

Dynamic Network Flow Models with Constant Attributes

First, "classic" dynamic network flow problems with constant attributes will be considered. The maximum dynamic s-t flow problem was presented by Ford and Fulkerson (1958). The goal is to send a maximum amount of flow from a source node s to a sink node t in a given time horizon. Ford and Fulkerson (1958) assumed the attributes of the problem, i.e. transit times and inflow capacities, to be constant. This problem can be easily interpreted in the evacuation context as evacuating as many as possible people from a danger zone (source node) into a safe zone (sink node). Ford and Fulkerson (1958) solve this problem optimally by letting start "temporally repeated flows" in every time period from the source such that the chain flows will reach the sink before the observed time horizon will end. The temporally repeated flows are based on flows computed with a certain minimum cost flow problem.

The earliest arrival flow problem, also called universal maximum flow problem, presented by Gale (1959) is an extension of the maximum dynamic s-t flow problem with an additional property: the cumulative amount of flow having reached the sink in every considered time period and all preceding time periods of the considered one have to be maximal. The earliest arrival flow translated in the evacuation context means that until every considered time period the maximal amount of evacuees enters the safe area. Considering, for example, an impending dam failure, this is a useful and reasonable property.

In case of multiple sources and/or sinks, additional parameters, e.g. supplies for sources and demands for sinks, and corresponding constraints may be added to the maximum dynamic flow problem or the earliest arrival flow problem. A maximum dynamic flow with the earliest arrival property does not necessarily exist if there are multiple sinks with predefined supplies and demands. Baumann and Skutella (2006) give an example with a single source and two sinks where no earliest arrival flow exists. For the evacuation scenario it is not a real limitation, if we assume that the exit, the sink, is not important in a case of an evacuation and the assignment of evacuees to exits is part of the decision making. This can be easily incorporated by adding a super sink to which all sinks are connected. The demand of the super sink is the sum of supplies of all sources, i.e. the total number of evacuees. Different variants of constraints that state, if the demand has to be satisfied within the time horizon or if as much as possible of the demand has to be satisfied are thinkable.

The quickest transshipment problem seeks to minimize the time the given supplies and demands in sources and sinks, respectively, are satisfied. For the evacuation scenario the quickest transshipment problem determines the evacuee-flow such that all evacuees leave the danger zone in a minimum amount of time or, in other words, the clearance time of the evacuation zone is minimized. The single source, single sink case of the quickest transshipment problem is called the quickest flow problem. The quickest path problem allows just a single path from a source to sink while minimizing the network clearance time (see Pascoal et al., 2006). The quickest transshipment problem with multiple sources and a single sink is called evacuation problem (see e.g. Hoppe and Tardos, 1994).

Burkard et al. (1993) use the relation between the quickest flow problem (single source, single sink) to the maximum dynamic network flow problem. They propose polynomial algorithms and a strongly polynomial algorithm to solve the quickest flow problem. The algorithms are based on parametric search with respect to the time horizon, like binary search. A maximum dynamic flow problem or an earliest arrival flow problem is solved several times with changing time horizon until a feasible solution regarding to the amount of flow that should reach the sink with the smallest time horizon is found.

Hoppe and Tardos (1994) present polynomial time algorithms for several dynamic network flow problems: the general multiple source quickest flow problem, the lexicographic maximum dynamic flow problem, and the earliest arrival problem. The lexicographic maximum dynamic flow problem maximizes amounts of flow leaving the sources in a specified order, i.e. the sources are ordered from high to low priority. They introduce "generalized temporally repeated flows", where flow is allowed to travel in the opposite direction of an edge. These algorithms are also explained in detail in Hoppe (1995) and a subsequent algorithm of the quickest transshipment problem is presented in Hoppe and Tardos (2000). Baumann and Skutella (2006) develop a strongly polynomial time algorithm to solve the earliest arrival flow problem with multiple sources and a single sink exactly, where the supplies and demand are given. Continuous dynamic flows are considered. Fleischer (2001) investigates the quickest transshipment problem, where the transit times of all arcs are zero, but the inflow is still assumed to be capacitated. Fleischer and Skutella (2007) present approximation algorithms for quickest multicommodity flows. A commodity is defined as a source–sink pair.

The minimum cost dynamic network flow problem seeks to find a dynamic network flow that satisfies the given supplies and demands such that the total cost defined e.g. on arcs or nodes is minimized while fulfilling the flow conservation and inflow capacity constraints. Minimum dynamic cost flow problems are considered in Chalmet et al. (1982), Hamacher and Tufekci (1987) and Hamacher and Tjandra (2001) for building evacuation. The average evacuation time per evacuee is minimized and waiting in nodes (storage in nodes) is allowed. Klinz and Woeginger (1995) and Klinz and Woeginger (2004) consider a dynamic minimum cost flow problem without waiting in nodes for the problem with a single source and a single sink. They state that there always exists a minimum cost flow, which does not use the opportunity to temporally store flow in nodes, if storage/waiting in nodes have nonnegative costs. Fleischer and Skutella (2002) prove the more general result that the minimum dynamic convex cost flow problem with multiple sources and sinks does not require storage in nodes. The supplies and demands that are assigned to sources and sinks have to be satisfied within the time horizon. The time when a certain amount of flow enters the network through a source, i.e. the network load, is not predetermined in advance.

If the time horizon is chosen high enough such that the flow assigned to the sources can leave the network within the given time horizon, then the "triple optimization result" (Jarvis and Ratliff, 1982) is valid. It states that an optimal solution of the dynamic maximization flow problem with the earliest arrival property or an optimal solution of the minimization of the weighted sum of flows entering the arcs pointing to the super sink solves the following three problems optimal: The earliest arrival problem, the minimization of the weighted sum and the quickest flow problem. The time copies of the sink are all connected to a super sink and the weights associated with the arcs that lead to the super sink are increasing with time.

Chalmet et al. (1982) refer to the result of Jarvis and Ratliff (1982) as the "triple optimization result". Jarvis and Ratliff (1982) proof the result for the single source, single sink case, but is also valid for the multiple source, single sink case (see e.g. Baumann and Skutella, 2006), where the sources have given supplies.

The result of Jarvis and Ratliff (1982) may be extended to the multiple sources, single sink case, but if the network capacities are to be determined in the optimization model then this result is no longer valid for the urban evacuation model we will present in the following chapters. We will give an example of an instance of our urban evacuation model, where the minimal cost flow leads to a longer clearance time and the minimum clearance time flow leads to a higher cost flow (see Sect. 3.5.1). In the urban evacuation model that we will introduce in Chap. 3, the inflow and the capacities of the arcs have to be determined and additionally a total capacity of the street segments will restrict the flow. A single sink situation is given for the urban evacuation model with the incorporation of the super sink and the fact that the street network sinks do not have a predetermined demand.

Dynamic Network Flows with Time-Dependent Attributes

The attributes of the dynamic flow problems were so far constant, but they may vary over time. The values of parameters varying over time are known in advance. This can be used to model the progress of a hazard like the rising of water when a dam failure occurs. Capacities may decrease over time or can be zero after a certain point in time. Dynamic network flow problems with time-varying attributes and solution methods is explicitly considered in the following literature: Tjandra (2003) considers time-dependent attributes, i.e. travel times, inflow capacities and costs. He presents solution algorithms for the single source, single sink case of the maximum dynamic network flow problem, the earliest arrival flow problem and the quickest flow problem with time-dependent attributes and time-dependent supplies as well as an algorithms solving time-dependent bicriteria dynamic shortest path problems (see also Hamacher et al., 2006).

A minimum cost dynamic flow problem on time-varying networks is considered by Cai et al. (2001) for the single source, single sink problem. Variables are the flow on an arc during a certain time period and the amount of flow waiting in a node during a certain time period. The time-varying attributes are the costs, travel times and capacities on arcs as well as on nodes. Algorithms with pseudopolynomial time complexity are suggested to solve the problem with waiting capacity, with unlimited waiting in nodes and with the prohibition of waiting in nodes. Miller-Hooks and Stock Patterson (2004) investigate an integral time-dependent quickest flow problem with time-varying attributes, i.e. arc travel times, node and arc capacities, and the supply at the source vary with time. They present an algorithm that solves the single source, single sink variant of the problem and, additionally, they propose a transformation of the network associated with the quickest transshipment problem into a single source, single sink problem such that the algorithm can be applied to the multiple source case, too. The demand is time-dependent, but the flow that has to meet a certain demand at a certain time may arrive in the sink before and wait there on a holdover arc of the super sink. Waiting in the super sink is not penalized with costs. Nasrabadi and Hashemi (2010) investigate a minimum cost flow problem in a discrete time model with time-varying transit times, transit costs, transit capacities, storage costs and storage capacities. An algorithm is presented to solve the considered problem.

Dynamic Network Flows with Flow-Dependent Travel Time

An extension of dynamic network flows such that affects of traffic flow are captured more accurately models the travel time dependent on flow, e.g. the travel time depends on the inflow of a certain point in time. With different models of flow-dependent travel times the following approaches try to map the traffic flow behavior more accurately while preventing computational tractability.

Kaufman et al. (1998) propose a mixed integer programming model for dynamic route guidance with traffic-dependent travel times that considers a discretization of an inflow-travel time relationship. The travel time depends on the amount of traffic inflow and is modeled with an alternative time-expanded network, where exactly one arc of finite different time-space arcs emerging from a time copy of a node has to be chosen. The traffic load onto the network is predetermined and may vary over time. A similar alternative time-expanded network is introduced by Carey (2001) to capture the flow behavior of traffic flow more realistically. They propose a linear model that is defined on an alternative time-expended network. The traffic flow is given a choice among time-space links. These links are associated with different travel times that depend on a given interval of inflow capacity. Thus the trip time of a vehicle on a link is influenced by the inflow rate when vehicles enter a link. Köhler et al. (2002) introduce two variants of an alternative time-expanded network that map a relaxation of inflow-dependent transit times similar to the approach of Carey (2001). The alternative time-expanded graphs are modeled in such way that standard network flow algorithms can be applied to solve the problems defined on them. The considered problem is the quickest dynamic s-t-flow problem. With an introduction to dynamic flows, these results are also presented in Langkau (2003). Köhler and Skutella (2005) introduce a model that considers load-dependent transit times. They assume that the total amount of flow traverses an arc at each point in time with uniform speed and that the speed depends only on the current amount of flow (or load) on that arc. An approximation algorithm for the quickest dynamic flow problem with load-dependent transit times is suggested. The earliest arrival flows, where the flow is defined according to the above-mentioned approaches of the inflow-dependent (Köhler et al., 2002) and load-dependent (Köhler and Skutella, 2005) travel times, are investigated in Baumann and Köhler (2007).

Another approach to capture the properties of traffic flow present Hall and Schilling (2005). They consider a rate-dependent flow model that takes the flow rate on any point on a link into account and present a heuristic for the quickest flow problem with rate-dependent travel time.

Surveys on Dynamic Network Flow Problems

For more details and different focuses on dynamic network flow problems, we refer to the following surveys: Aronson (1989) gives a survey on discrete dynamic network flows, especially on the maximum dynamic flow and the minimum dynamic flow problem, on corresponding solution techniques and he lists an extensive number of references of examples and applications. Mentioned examples are dynamic traffic assignment models or problems that occur in communication or traffic systems. Hamacher and Tjandra (2001) give an overview over mathematical modeling of evacuation problems concentrating mainly on building evacuation. They especially present variations of discrete time dynamic network flow problems that can be used to model evacuation problems, like the minimum cost dynamic network flow, earliest arrival flow or quickest flow problems and as well as solution algorithm of the corresponding problems. Kotnyek (2003) gives an overview of dynamic network flows. He concentrates on the "basic" dynamic problems, the minimum dynamic cost flow, the maximum dynamic flow, the earliest arrival flow and quickest flow problem as well as on generic solution techniques. Skutella (2008) presents a survey of continuous time dynamic network flows with constant travel times and capacities. He concentrates on the maximum s-t flow problem and earliest arrival problem with a single sink s and a single source t and solution algorithms. In "Traffic Networks and Flows over Time", Köhler et al. (2009) summarize approaches and algorithms of static flows and dynamic flows with and without the incorporation of congestion. They review dynamic networks flows with constant and especially flow-dependent travel times.

Evacuation Models Based on Dynamic Network Flow

Explicitly for the evacuation of pedestrians developed dynamic network flow models are e.g. for building evacuation (Chalmet et al. 1982, Choi et al. 1988 or Chen and Miller-Hooks 2008). Chalmet et al. (1982) model a dynamic network flow problem that minimizes the average time an evacuee needs to exit the building. Using the "triple optimization result" proven by Jarvis and Ratliff (1982) the considered problem solves also the maximization of the total number of evacuees as well as the minimization of the total evacuation time. Hamacher and Tufekci (1987) consider among other variants, a building evacuation problem that avoids unnecessary movements during evacuation. Both evacuation models, i.e. the models presented in Chalmet et al. (1982) and Hamacher and Tufekci (1987), allow multiple sources and define a single super sink to which the exits of the buildings are connected in the constructed network. Choi et al. (1988) model different dynamic network flow problems for buildings that take flow-dependent capacity into account. For special graphs, so called path structured networks and tree structured networks, which can be considered as typical for common buildings, greedy algorithms are presented. Chen and Miller-Hooks (2008) incorporate "shared information" (modeled with binary variables on arcs and in every point in time) in their building evacuation problem that updates the evacuation routes in every point in time. The travel times and arc inflow capacities are time-dependent and known a priori. Evacuations of people from hospitals (e.g. Taaffe et al., 2005) or naval ships (e.g. Pérez-Villalonga et al., 2008) are also investigated.

Mamada et al. (2005) consider an evacuation problem on tree dynamic networks with the property "all the supplies going through a common vertex are sent out through a single arc incident to it toward a single sink" (Mamada et al., 2005, p. 196). That means the flow is not allowed to diverge on different arcs. They consider a continuous time quickest transshipment problem.

For evacuation from regions, the following approaches which optimize traffic flows that have to leave the network due to a specific objective have been developed: Yamada (1996) investigates shortest path and the minimal cost flow problem on static networks for city emergency evacuation. Lu et al. (2005) propose a heuristic approach to solve the problem of the minimization of evacuation egress time with time-dependent node and arc capacity for large-scale instances. The heuristic algorithm calculates evacuation plans that provide evacuation routes and evacuation schedules without using the time-expanded network, i.e. using the static network. Kim et al. (2007) improve the heuristic proposed in Lu et al. (2005) in terms of the runtime using the min-cut max-flow theorem. Time-dependent travel times and capacities are allowed. Lim et al. (2009) apply the maximum dynamic network flow problem for the case of hurricane evacuation to determine an upper bound of the total amount of evacuees that reach a safe zone within a given time horizon. Analog to the approach of Burkard et al. (1993) for the single source and sink case, a binary search is executed until the minimum time horizon is found such that the model is feasible. Kamiyama et al. (2006) suggest an algorithm for an evacuation problem modeled as a quickest flow problem with single sink. The algorithm is constructed for a grid network with uniform inflow capacity.

Evacuation Models Based on Dynamic Traffic Assignment Problems

A simple traffic assignment model defined on a static network is used for a regional evacuation approach in Han et al. (2006a,b). The assignment model is based on a system optimum traffic assignment model of Sheffi (1984). Amounts of flow of the demand of sources is assigned to routes leading to a super sink with the objective to minimize the total travel cost for all travelers in the network. Ng and Waller (2009) suggest a slightly different problem. They consider a two-stage stochastic programming network design model for transportation networks that take all possible evacuation scenarios into account. Capacity extensions for the transportation network of a certain region are determined in the first stage. In the second stage, a system optimal static traffic assignment problem is considered to achieve a solution for the evacuation traffic flow. The trade-off of the minimization of the total evacuation time, total distance and traffic congestion is intended by Stepanov and MacGregor Smith (2009) with a multiobjective evacuation routing model on the static network that considers the demand of the origin nodes and the capacities of the destinations as well as of the links. Exactly one route is assigned out of a set of k-shortest paths to an origin-destination pair such that the weighted sum of the three objectives is minimized.

Ziliaskopoulos (2000) developed a linear model for a system optimum dynamic traffic assignment problem with a single destination that is based on the cell transmission model of Daganzo (1994, 1995). The cell transmission model (Daganzo, 1994, 1995) is a discrete approximation of the Lighthill, Whitham, Richards hydrodymatic model used to model traffic evolution. The street network is divided into cells such that in one period a vehicle can travel trough the cell in light traffic. The time horizon is divided into time periods of equal length. These cells are connected with connectors through which flow can change a cell. Every period, the network situation is updated regarding flow conservation, inflow capacity in and out of cells and total capacity within cells. The travel time is implicitly modeled in the construction of the cells and the length of the period. It is not explicitly used like, for example, in the flow conservation constraint of a dynamic network flow model. The model of Ziliaskopoulos (2000) itself builds the basis for some evacuation models in the regional environment: The linear optimization program presented in Chiu et al. (2007) models a mass evacuation problem using a dynamic traffic flow optimization model that is based on the formulation of Ziliaskopoulos (2000). The loading pattern is not given a priori, it is determined with the model. Liu et al. (2006b) model a staged evacuation process on a network with a predefined evacuation zones, where the starting time to begin the evacuation of each zone has to be determined. With the start of the evacuation in a certain zone, the predefined loading pattern loads the flow onto the network of an evacuation zone. They present in Liu et al. (2006c) a revised cell transmission model that is designed for large-scale networks. This model uses a revised flow propagation formulation and allows different cell sizes. Liu et al. (2006c) propose two evacuation models: a model that maximizes the number of evacuees reaching the sinks within a given time horizon and a model that minimizes the total time a given amount of evacuees are within the evacuation network, respectively.

Uncertainty aspects are incorporated in the following cell transmission framework: Road capacities may be uncertain in an evacuation situation. The capacities may decrease for example by flooding. Yazici and Ozbay (2007) take link capacity changes into account by introducing probabilistic capacity constraints. Additionally, they use the solution of the model to determine the capacity of shelters that are associated with the given sinks. Ng and Waller (2010) suggest an evacuation route planning model that takes the uncertainty of demand as well as the uncertainty of road capacity into account. The parameter values of a cell transmission-based evacuation model are adjusted such that a more conservative plan is generated than without the manipulation of the values. The values of the demand are increased and the capacity values are decreased subject to certain assumptions. A robust optimization approach for evacuation planning is developed by Yao et al. (2009). They focus on demand uncertainty. A "threat level" weight is introduced into the objective function that is used to penalize a solution where evacuees are still in the evacuation area at the end of the time horizon.

In all above-presented approaches, the capacities are known in advance. That means, the evacuation routes or traffic routing for the case of the evacuation of a region is already determined. In Sect. 2.2, we will give a survey on optimization models that incorporate traffic management strategies like the temporally reversal of lane in the case of an evacuation.

2.2 Evacuation Models with Traffic Management Strategies

Traffic management strategies that are used for evacuation planning like lane-reversal operations are modeled with optimization models. The temporary reversal of lanes of certain street segments are also known in rush-hour traffic. Zhang and Gao (2007) present for example a bi-level programming model that reallocates lanes for rush-hour traffic.

Traffic conflicts within intersections are the crossing of lanes and the merging of lanes. Cova and Johnson (2003) developed a mixed integer programming model that minimizes the total travel distance while forbidding intersection crossing conflicts and allowing a fixed number of merging conflicts. The network is constructed lane-based, i.e. each lane of every considered street segment, also every lane within intersections, are depicted by an arc. The intersection subgraphs consist of nodes that represent the access of every lane. That means, if four lanes access an entrance of an intersection then four nodes are generated for this entrance of the considered intersection.

Certain network flow problems with the goal to determine an optimal arc reversal are NP-hard. Rebennack et al. (2010) investigated the complexity of network flow problems, where the possibility of arc reversal is given. They show that the maximum dynamic flow problem with arc reversal capability having a single source and a single sink is polynomial solvable. In contrast, the maximum dynamic network flow problem with arc reversal with only an additional source or sink turns out to be NP-complete. Moreover, they show that the dynamic transshipment contraflow problem is NP-complete, even with only two sources and one sink or only one source and two sinks. A directed network with sets of sources and sinks, capacities and symmetric travel times on arcs is given. Then the dynamic transhipment contraflow problem asks, if there exists a feasible flow within a given time bound T, where it is allowed that each arc is reversed once at time 0. The dynamic network flow problem of minimizing the total cost resulting from arc switching costs proves to be NP-hard. Kim and Shekhar (2005) consider a similar dynamic contraflow problem, where a network with sets of sources and sinks is given (each source has a initial occupancy) and each arc has a capacity and a travel time. The contraflow problem asks, if there exists a fixed contraflow network configuration such that the minimal evacuation time is lesser or equal to a given time bound T. They give a sketch of a proof that the contraflow problem is NP-complete.

The following models restructure the traffic routing for the case of an evacuation with traffic management strategies like lane-reversal, addition of lanes or the consideration of traffic conflicts and describe the traffic flow over time.

Tuydes and Ziliaskopoulos (2006) propose a linear optimization model based on the cell transmission model for traffic flow that includes the capability of the reversal of lanes for the case of evacuation. The decision variable corresponding to the number of lanes is assumed to be continuous. The proposed solution method is a tabu search-based heuristic algorithm incorporating results of simulation.

A linear cell transmission-based optimization model is also investigated by Kimms and Maassen (2011b). The optimization model incorporates the level of danger of different parts of the network in the objective such that it is more preferable to use streets with a lower "danger" level. The direction of lanes is assigned afterwards depending on the result of flow using a cell. Kimms and Maassen (2011a) extend their model introducing different cell sizes as well as incorporating continuous decision variables representing the numbers of lanes. The goal is to achieve an optimized traffic routing consisting of one-way streets. Crossing conflicts, "touching" conflicts and the merging of lanes are prohibited, and the number of turning directions within an intersection that can be used by traffic are bounded. "Touching" conflicts are understood as conflicts that may be caused due to a limited amount of space within intersections (see for a detailed explanation Kimms and Maassen 2010b): Consider the four directions that are generated if two disjunct pairs of entrances of an intersection are chosen and each of the pairs generate two directions. Then exactly one direction out of the four can be traveled by flow to avoid a "touching" conflict. Kimms and Maassen (2010a) present heuristic approaches that are based on a shortest path approach and the computation of a feasible solution based on the solution of an auxiliary model defined on the static network.

The cell transmission-based mixed integer programming model of Peeta and Kalafatas (2008) gives the opportunity that predefined network design options like lane-reversal, lane addition on certain street segments are chosen and prohibits crossing conflicts while minimizing the cumulated sum of total time spent in the network. The network load (demand per time interval) is given, the costs of network design options are included and the total costs of implementing the network design options are restricted by a given bound. The costs of implementing e.g. a lane-reversal operation are understood as the cost of special equipment like electronic variable signage or the cost of training the personnel. They (Peeta and Kalafatas, 2009) propose a similar model considering just lane-reversal operations and investigate a test network varying the values of the parameters population size, number of contraflow operations and the distribution of the origin-destination demand.

Xie et al. (2010) and Xie and Turnquist (2011) introduce a lane-based evacuation problem with a bi-level model that determines a reallocated fixed traffic routing for the evacuation case. The capability of lane-reversal is integrated, crossing conflicts are prohibited and merging of lanes is allowed. The traffic management operations are presented for the case of two lanes for each street segment. The upper level problem seeks in both cases to optimize the system-wide evacuation performance. The lower level problem assigns traffic flow to the corresponding network components building a user-equilibrium. In Xie et al. (2010) the model describing the traffic flow is realized with a mixed integer linear cell transmission-based model, in Xie and Turnquist (2011) the traffic flow is described by a mixed integer linear dynamic network flow problem. For both approaches a corresponding integrated Lagrangian relaxation and tabu search approach is suggested.

Kim and Shekhar (2005) solve a quickest transshipment problem with lane-reversal heuristically. They propose heuristic approaches: a greedy algorithm that assigns lanes depending on the optimal values of flow of a minimum cost flow defined on the time-expanded network, a heuristic based on simulated annealing metaheuristic and a bottleneck heuristic (Kim et al., 2008). The basic idea of the latter one, is to identify bottlenecks and increase its capacity by lane-reversal iteratively.

2.3   Further Problems in Evacuation Planning

In an evacuation scenario, there are not just the evacuees involved that travel by vehicle to the exits of the evacuation zone, but there are also pedestrians and also emergency response vehicles that have to travel to the evacuation zone and within the evacuation zone in counter direction to the evacuees to reach places where they are needed. These are services like the transportation of emergency equipment into the evacuation zone or buses and emergency ambulances that support evacuees that are not able to self evacuate. The buses and ambulances may have to travel multiple time into and out of the evacuation zone.

Liu et al. (2006a) present an approach for the evacuation of pedestrians caused by flood. The objective is to determine the evacuation route to minimize the evacuation time taking the time-varying travel time on arcs into account. In the case of a flood disaster the water depth varies over time and influences the walking speed of people.

Chiu and Zheng (2007) study a cell transmission-based dynamic traffic assignment model that models different groups of evacuees and emergency response vehicles. A priority factor is included in the objective that minimizes the priority weighted sum of total travel times of vehicles to the corresponding destination. Xie and Turnquist (2009) consider in the presented evacuation problem flow of evacuees and flow of emergency vehicles. They propose an optimization model with bi-level structure that needs the number of lanes assigned for the emergency vehicles in advance and determines the traffic routing for the evacuation flow on a static network. Crossing conflicts are prohibited and the reversal of lanes is allowed. But they do not discuss, if there exists a feasible solution without crossing conflicts for every network and demand pattern of evacuees and emergency vehicles. Indeed there are simple examples, where no feasible solution without a crossing of two lanes exists. Such an example is presented in Chap. 7. A similar problem consider Kimms and Maassen (2010b). They incorporate the flow of rescue teams that may travel to a place within the evacuation area or commute between a safe and an endangered place. A street segment can be either used for evacuation traffic or for rescue traffic in different time periods. I.e. in certain time periods street segments have to be closed for evacuation traffic. The assignment of lanes is static, but the permission of flow to use the corresponding street section may vary over time. The rules of traffic routing are the same as introduced in Kimms and Maassen (2010a) and have to be valid for the rescue flow and evacuation flow separately.

There are also multicommodity problems that provide in the same time evac-uation planning activities as well as disaster response activities: Özdamar and Yi (2007) suggest a location-distribution model for the coordination of logistics support and evacuations in disaster response activities. Here, evacuation is understood as the evacuation and transfer of wounded people to emergency units. The distribution of commodities like medical materials, medical personnel or food to locations in the affected area, the transportation of wounded people to hospitals or emergency centers and the determination of temporary emergency centers is formulated in a mixed integer capacitated location-routing model. It minimizes the delay in satisfying demands of prioritized commodities. In Özdamar and Yi (2008), they present the problem of the evacuation from affected areas to medical centers during the initial response phase and logistic transport from supply centers to distribution centers in affected areas and they suggest a heuristic based on greedy neighborhood search. Also Osman et al. (2009) incorporate the evacuation problem in a disaster response problem. They present a general transportation routing problem for disaster areas that considers the transportation of materials and people out of a disaster area and the transportation of materials and people into the disaster area.

Services like the amount of provided beds of shelters are restricted. Hence, the decision what shelters in the safe zone are operated may be incorporated in the formulation of traffic assignment problems for evacuation scenarios. A location-allocation model that captures both, the choice of shelters and the network flow problem, is presented by Sherali et al. (1991). They propose a non-linear mixed-integer programming problem that takes the available shelter staff into account and describes the evacuation flow on a static network, where flow is assigned to capac-itated links from sources to destinations. Kongsomsaksakul et al. (2005) propose a location-allocation model for evacuations planning with a bi-level programming formulation. With the upper level the locations of shelters are chosen among a given set, and the lower level determines the allocation of sources to shelters and assigns the traffic of the evacuees to routes to the corresponding shelter. A genetic algorithm is presented to solve the proposed model heuristically. Saadatseresht et al.

(2009) suggest a multiobjective optimization problem that assigns building blocks to safe areas regarding the distance, the population size of the building block and the capacity of the corresponding shelters in the safe areas.

Besides the reallocation of people to safer locations due to a disaster, people may need help of emergency services of ambulances or of firefighters. During a disaster some of the facilities like ambulances or fire stations may be damaged. Huang et al. (2010) consider the problem of choosing locations for emergency service facilities taking large-scale disasters into account, where some facilities may become inoperable. They propose a variation of the p-center problem.

Another problem is the decision if to order an evacuation. For example, Regnier (2008) investigates the problems of evacuation decisions concerning whole regions or cities due to hurricanes. Evacuation decisions because of the risk of hurricanes are based on uncertain data. The forecasts cannot predict with 100% certainty if a hurricane will strike a special region. False evacuation alarms can cause costs like the costs for preparation or the "cost" of non compliance of people in the future. Therefore a trade-off between the danger to residents and the costs of false evacuation has to be made. To improve the quality of these decisions, the uncertainty of the forecasts can be measured and incorporated in the decision process. To support the evacuation decision, Regnier (2008) develops a decision support model that estimates the uncertainty of the forecasts based on historic hurricane tracks.

Cova and Church (1997) consider a problem that evaluates the maximal "vulnerability" of a community in potential evacuation regions that are endangered by moving hazards with a high level of uncertainty. As examples for that kind of hazards urban firestorms or toxic spills on highways are mentioned. The vulnerability of a node representing an intersection in the considered network is described by the average population per exit lane. The developed model is a nonlinear binary optimization problem and has to be applied for every node in the network. To solve the problem for arbitrary large networks in adequate time, Cova and Church (1997) propose a heuristic algorithm.

2.4 Simulation Approaches

Simulations are used to evaluate given traffic networks under a certain or under different scenarios. Simulation methods are also applied to decision support in evacuation planning. In the considered traffic networks, bottlenecks are identified and the evacuation time is estimated. Based on these results, traffic operational strategies can be implemented. Macroscopic simulation models for regional evacuation planning are for example NETVACI (Sheffi et al., 1982) or MASSVAC (Hobeika and Jamei, 1985) and microscopic simulation models are for example CORSIM (e.g. used in Theodoulou and Wolshon, 2004) or VISSIM (e.g. used in Edara et al., 2010). More simulation models, macroscopic as well as microscopic models, that are constructed for regional evacuation are listed for example in Pel et al. (2010) or Edara et al. (2010).

Frameworks for regional evacuation modeling are reviewed and proposed e.g. by Southworth (1991) and Barrett et al. (2000). Southworth (1991) considers a framework of regional evacuation modeling as a five step approach: traffic generation, traffic departure times, destination selection, traffic route selection and implementation of traffic management controls. He summarizes necessary data like the estimation of population, the estimation of the number of vehicles or the determination of the traffic loading curve. Difficulties arise for example for the estimation of day time population.

Barrett et al. (2000) consider an evacuation modeling framework for hurricane evacuation to determine short range strategies during the storm. A goal is to determine real time operational strategies and to develop a model that can analyze large regional networks in real time that incorporates traffic conditions based for example on the data of road sensors or law enforcement personnel.

Simulation tools provide generally various possibilities of alternative scenarios like different weather conditions or lane management strategies (see e.g. Sheffi et al. 1982, or the "knowledge based mode" in TEDSS, Hobeika et al. 1994). Measures are e.g. the estimated evacuation time (network clearance time) or congestion on links. Components of traffic simulation models are generally the network loading, trip generation, traffic assignment and the update of the traffic condition like travel times. These components are iteratively operated until all vehicles have left the network (see e.g. Hobeika et al., 1994).

There are different approaches of simulation tools for different disasters like nuclear accidents, hurricanes, floods or wild-fires:

The simulation approach of Sheffi et al. (1982) was motivated by the requirement of estimating the clearance time for areas surrounding sites of nuclear power plants. Sheffi et al. (1982) provide a macroscopic traffic simulation model called NETVACI to simulate traffic patterns during a mass evacuation of an area. Also Hobeika et al. (1994) present a decision support system TEDSS (transportation evacuation decision support system) for the development of evacuation plan around nuclear power plants. The proposed model is based on the simulation tool MASSVAC that simulates the dynamic movement of traffic on the highway network. The proposed evacuation policy is that evacuees have to leave the danger zone (10 mile zone) first before they can drive to the proper shelter somewhere in the safe area. Traffic management strategies like the permission to use shoulder lanes or lane-reversal operations may be implemented in a scenario and may be evaluated. An update of the above introduced tool is presented in Hobeika and Kim (1998). An user equilibrium assignment algorithm is incorporated. The user equilibrium requires that the travel times are equal on all used paths for each origin-destination pair and that no other path leads to a lesser travel time.

Southworth and Chin (1987) use MASSVAC to investigate alternative traffic routings for the case of an emergency evacuation due to flooding that results of a dam failure. Pel et al. (2010) consider time-varying networks that map the progress of a moving hazard. They investigate evacuation scenarios due to flooding with the simulation model EVAQ in a case study of the Dutch city Rotterdam.

Farahmand (1997) presents a simulation model applied for hurricane evacuation planning in the US-state of Texas in the Rio Grande Valley. Different scenarios depending on hurricane categories are investigated. Williams et al. (2007), Theodoulou and Wolshon (2004) and Edara et al. (2010) use microscopic simulation models to evaluate freeway networks for the case of a hurricane evacuation in North Carolina, Louisiana and Virginia, respectively. Traffic management strategies like lane-reversals are taken into account and scenarios depending e.g. on storm categories are investigated. Edara et al. (2010) considers especially large-scale hurricane networks.

Cova and Johnson (2002) use microscopic traffic simulation to study evacuation plans in the urban-wildland interface. An urban-wildland interface is an area "where urban growth encroaches into fire-prone wildlands" (Cova and Johnson, 2002, p. 2211). This approach is applied to the Emigration Canyon in Utah for different scenarios in terms of the number of vehicles per household and the possibility of the usage of second access road.

Different evacuation strategies are investigated for different road patterns by Chen and Zhan (2008). They study the performance of simultaneous and staged evacuation strategies using agent-based simulation. For the staged strategies the evacuation area is divided into zones, and these zones are evacuated in different sequences. The evacuation strategies, the simultaneous and all possible staged strategies, are applied to three different types of road network structures, i.e. a grid road structure, a ring road structure and a real road structure from the City of San Marcos (Mexico). The results of the study of Chen and Zhan (2008) are that no evacuation strategy is the most effective strategy across the analyzed network structures and that the effectiveness measured by the overall evacuation time of a strategy depends on the network structure and the density of the population.

2.5 Summary

The literature review indicates that evacuation planning is an important research topic. Simulation approaches are developed for various situations like hurricanes, floods or nuclear accidents that may cause the need to evacuate a certain area (compare Sect. 2.4). These hazards may endanger a population and an evacuation may be the best choice to protect the affected population.

Dynamic network flow models that describe evacuation problems like the reallocation of evacuees to a safe zone in the minimum amount of time (the quickest flow problem) or the reallocation of the maximal amount of evacuees in a given amount of time (maximum dynamic network flow problem) are studied and investigated with different properties of the attributes. The optimization models are applied to evacuation planning for buildings as well as regions with time- and flow-dependent attributes. Dynamic traffic assignment models are applied to optimize evacuations from regions incorporating also aspects of uncertainty. The evacuation of a region may be the evacuation of a district of a city, a whole city or may affect multiple cities. All these problems and models (presented in Sect. 2.1) consider the capacities of the arcs, and hence the traffic routing, as fixed. The traffic routing of the considered street network is known and is not part of the optimization.

With the incorporation of the possibility of lane-reversal operations for street networks, traffic routing becomes part of the optimization. Scarce capacity has to be allocated to the two different directions of street sections or, if only one-way streets are intended, the capacity has to be allocated to exactly one of the two directions. Additional decision variables are necessary to model the traffic routing for example with integer number of lanes or indicator variables that determine which direction of a street segment is permitted to be traveled by traffic flow. Dynamic network flow problems become harder to solve if there are more than one source or more than one sink.

Studies on the optimization of traffic networks that intend to optimize the traffic routing with the capability of lane-reversal and the prohibition of crossing conflicts are rare. There are recently Kimms and Maassen (2010a,b, 2011a), Peeta and Kalafatas (2009), Xie et al. (2010) as well as Xie and Turnquist (2011). Each of them considers the progress of traffic flow over a certain time horizon, whereas each of them takes different definitions of crossing conflicts into account. Peeta and Kalafatas (2009) only mention that the crossing of flows is prohibited. Kimms and Maassen (2011a) consider one-way streets and allow a limited number of chosen directions within intersections. Xie et al. (2010) and Xie and Turnquist (2011) lean upon the definition given in Cova and Johnson (2003). They present lane-based models for only two lanes per street segment and intersections with four entrances/exits.

The work of Bretschneider and Kimms (2011a,b) can be cited in Sect. 2.2 (Evacuation Models with Traffic Management Strategies). The traffic re-routing measures lane-reversal, certain restrictions of merging of lanes and the measure prohibition of crossing conflicts are considered in dynamic network flow models. In this work and in the articles Bretschneider and Kimms (2011a,b) which build the basis of parts of this work, a direction-based network will be investigated. I.e. not every lane is modeled in the network like in Xie et al. (2010). But the idea of crossing conflicts is based on the lane-based definition given in Cova and Johnson (2003), too. The lanes within the network can be allocated such that no crossing conflict occurs if the assigned number of lanes is feasible with respect to the proposed constraints. A notation for the network will be introduced such that all crossing conflicts can be determined generally for all intersections such that the determination of directions which cross is comprehensible. Contraflow operations that allow the partly reversal of lanes and contraflow operations that lead to one-way streets are studied which have an effect on the number of crossing conflicts.

References

Aronson JE (1989) A survey of dynamic network flows. Ann Oper Res 20:1–66
Barrett B, Ran B, Pillai R (2000) Developing a dynamic traffic management modeling framework for hurricane evacuation. Transport Res Rec: J Transport Res Board 1733:115–121
Baumann N, Köhler E (2007) Approximating earliest arrival flows with flow-dependent transit times. Discrete Appl Math 155(2):161–171

Baumann N, Skutella M (2006) Solving evacuation problems efficiently–earliest arrival flows with multiple sources. In: 47th annual IEEE symposium on foundations of computer science (FOCS'06), pp 399–410

Bretschneider S, Kimms A (2011a) A basic mathematical model for evacuation problems in urban areas. Transport Res A Pol Pract 45:523–539

Bretschneider S, Kimms A (2011b) Pattern-based evacuation planning for urban areas. Eur J Oper Res 216:57–69

Burkard R, Dlaska K, Klinz B (1993) The quickest flow problem. ZOR Meth Models Oper Res 37:31–58

Cai X, Sha D, Wong CK (2001) Time-varying minimum cost flow problems. Eur J Oper Res 131(2):352–374

Carey M (2001) Dynamic traffic assignment with more flexible modelling within links. Network Spatial Econ 1:349–375

Chalmet LG, Francis RL, Saunders PB (1982) Network models for building evacuation. Manag Sci 28(1):86–105

Chen L, Miller-Hooks E (2008) The building evacuation problem with shared information. Nav Res Logist 55:363–376

Chen X, Zhan F (2008) Agent-based modelling and simulation of urban evacuation: relative effectiveness of simultaneous and staged evacuation strategies. J Oper Res Soc 59:25–33

Chiu Y-C, Zheng H (2007) Real-time mobilization decisions for multi-priority emergency response resources and evacuation groups: model formulation and solution. Transport Res Part E: Logist Transport Rev 43(6):710–736, Challenges of Emergency Logistics Management

Chiu Y-C, Zheng H, Villalobos J, Gautam B (2007) Modeling no-notice mass evacuation using a dynamic traffic flow optimization model. IEE Trans 39:83–94

Choi W, Hamacher HW, Tufekci S (1988) Modeling of building evacuation problems by network flows with side constraints. Eur J Oper Res 35:98–110

Cova TJ, Church RL (1997) Modeling community evacuation vulnerability using GIS. Int J Geograp Inform Sci 11(8):763–784

Cova TJ, Johnson JP (2002) Microsimulation of neighborhood evacuations in the urban-wildland interface. Environ Plann A 34:2211–2229

Cova TJ, Johnson JP (2003) A network flow model for lane-based evacuation routing. Transport Res A Pol Pract 37(7):579–604

Daganzo CF (1994) The cell transmission model: a dynamic representation of highway traffic consistent with the hydrodynamic theory. Transport Res B 28B(4):269–287

Daganzo CF (1995) The cell transmission model Part II: network traffic. Transport Res B 29B(2):79–93

Edara P, Sharma S, McGhee C (2010) Development of a large-scale traffic simulation model for hurricane evacuation—methodology and lessons learned. Nat Hazards Rev 11(4):127–139

Farahmand K (1997) Application of simulation modeling to emergency population evacuation. In: Proceedings of the 1997 winter simulation conference, ACM, Atlanta, GA, 7–10 Dec 1997, pp 1181–1187

Fleischer L (2001) Faster algorithms for the quickest transshipment problem. SIAM J Optim 12:18–35

Fleischer L, Skutella M (2002) Minimum cost flows over time without intermediate storage. In: Proceedings of the 14th annual ACM-SIAM symposium on discrete algorithms, Baltimore, Maryland Publisher Society for Industrial and Applied Mathematics Philadelphia, PA, 12–14 Jan 2003, pp 66–75, 2002

Fleischer L, Skutella M (2007) Quickest flows over time. SIAM J Comput 36(6):1600–1630

Ford LR Jr, Fulkerson DR (1958) Constructing maximal dynamic flows from static flows. Oper Res 6(3):419–433

Gale D (1959) Transient flows in networks. Michigan Math J 6(1):59–63

Hall A, Schilling H (2005) Flows over time: towards a more realistic and computationally tractable model. In: Proceedings of the 7th workshop on algorithm engineering and experiments (ALENEX05), SIAM, pp 55–67, 2005

Hamacher H, Tjandra S (2001) Mathematical modeling of evacuation problems: a state of art. Berichte des Frauenhofer ITWM, Nr. 24. http://www.itwm.fraunhofer.de/fileadmin/ITWM-Media/Zentral/Pdf/Berichte_ITWM/2001/bericht24.pdf. Accessed 18 Oct 2006

Hamacher HW, Tufekci S (1987) On the use of lexicographic min cost flows in evacuation modeling. Nav Res Logist 34:487–503

Hamacher H, Ruzika S, Tjandra S (2006) Algorithms for time-dependent bicriteria shortest path problems. Discrete Optim 3:238–254

Han LD, Yuan F, Chin S, Hwang H (2006a) Global optimization of emergency assignments. Interfaces 36(6):502–513

Han LD, Yuan F, Chin S, Hwang H (2006b) Proposed framework for simultaneous optimization of evacuation traffic destination and route assignment. Transport Res Rec: J Transport Res Board 1964:50–58

Hobeika A, Kim C (1998) Comparison of traffic assignments in evacuation modeling. IEEE Trans Eng Manag 45(72):192–198

Hobeika A, Kim S, Beckwith RE (1994) A decision support system for developing evacuation plans around nuclear power stations. Interfaces 24(4):22–35

Hobeika AG, Jamei B (1985) MASSVAC: a model for calculating evacuation times under natural disasters. In: Proceedings on the conference on emergency planning, Simulation Series 15, vol 5, pp 23–28

Hoppe B (1995) Efficient dynamic network flow algorithms. Ph.D. thesis, Cornell University, Ithaca, NY

Hoppe B, Tardos E (1994) Polynomial time algorithms for some evacuation problems. In: Proceedings of the fifth annual ACM-SIAM symposium on discrete algorithms 1994, Publisher Society for Industrial and Applied Mathematics Philadelphia, PA, pp 433–441, 1994

Hoppe B, Tardos E (2000) The quickest transshipment problem. Math Oper Res 25(1):36–62

Huang R, Kim S, Menezes MBC (2010) Facility location for large-scale emergencies. Ann Oper Res 181:271–286

Jarvis JJ, Ratliff HD (1982) Some equivalent objectives for dynamic network flow problems. Manag Sci 28(1):106–108

Kamiyama N, Katoh N, Takizawa A (2006) An efficient algorithm for evacuation problems in dynamic network flows with uniform arc capacity. In: Cheng S-W, Poon C (eds) Algorithmic aspects in information and management. Lecture notes in computer science, vol 4041. Springer, Berlin, pp 231–242

Kaufman DE, Nonis J, Smith RL (1998) A mixed integer linear programming model for dynamic route guidance. Transport Res B 32(6):431–440

Kim S, Shekhar S (2005) Contraflow network reconfiguration for evacuation planning: a summary of results. In: Proceedings of the 13th annual ACM international workshop on geographic information systems. GIS '05. ACM, New York, pp 250–259

Kim S, George B, Shekhar S (2007) Evacuation route planning: scalable heuristics. In: Proceedings of the 15th annual ACM international symposium on advances in geographic information systems. GIS '07. ACM, New York, pp 20:1–20:8

Kim S, Shekhar S, Min M (2008) Contraflow transportation network reconfiguration for evacuation route planning. IEEE Transact Knowl Data Eng 20(8):1115–1129

Kimms A, Maassen K-C (2010a) A fast heuristic approach for large scale cell-transmission-based evacuation planning. Working paper, University of Duisburg-Essen

Kimms A, Maassen K-C (2010b) Cell-transmission-based evacuation planning with rescue teams. Working paper, University of Duisburg-Essen

Kimms A, Maassen K-C (2011a) Extended cell-transmission-based evacuation planning in urban areas. Pesquisa Operacional 31:405–441

Kimms A, Maassen K-C (2011b) Optimization and simulation of traffic flows in the case of evacuating urban areas. OR Spect 33:571–593

Klinz B, Woeginger GJ (1995) Minimum cost dynamic flows: the series-parallel case. In: Proceedings of the 4th international IPCO conference on integer programming and combinatorial optimization. Springer, London, pp 329–343

Klinz B, Woeginger GJ (2004) Minimum cost dynamic flows: the series-parallel case. Networks 43(3):153–162

Köhler E, Skutella M (2005) Flows over time with load-dependent transit times. SIAM J Opt 15(4):1185–1202

Köhler E, Langkau K, Skutella M (2002) Time-expanded graphs for flow-dependent transit times. In: Mhring R, Raman R (eds) Algorithms – ESA 2002. Lecture notes in computer science, vol 2461. Springer, Berlin, pp 49–56

Köhler E, Möhring RH, Skutella M (2009) Traffic networks and flows over time. In: Lerner J, Wagner D, Zweig K (eds) Algorithmics of large and complex networks. Lecture notes in computer science, vol 5515. Springer, Berlin, pp 166–196

Kongsomsaksakul S, Yang C, Chen A (2005) Shelter location-allocation model for flood evacuation planning. J E Asia Soc Transport Stud 6:4237–4252

Kotnyek B (2003) An annotated overview of dynamic network flows. Technical report, INRIA, Paris, Nr. 4936. http://hal.inria.fr/inria-00071643/

Langkau K (2003) Flows over time with flow-dependent transit times. Ph.D. thesis, Technische Universität Berlin, Germany

Lim GJ, Zangeneh S, Baharnemati MR, Assavapokee T (2009) A simple binary search algorithm for short notice evacuation scheduling and routing. In: Proceedings of the 2009 industrial engineering research conference. http://www.serri.org/publications/Documents/JSU%20Project%2070004%20IERC2009%20-%20Technical%20Paper%20by%20UH%20%28Aliabadi%29.pdf, 18 December 2010, 10:28 MEZ

Liu Y, Hatayama M, Okada N (2006a) Development of an adaptive evacuation route algorithm under flood disaster. Ann Disast Prev Res Inst. Kyoto University 49:189–195

Liu Y, Lai X, Chang G-L (2006b) Cell-based network optimization model for staged evacuation planning under emergencies. Transport Res Rec: J Transport Res Board 1964:127–135

Liu Y, Lai X, Chang G-L (2006c) Two-level integrated optimization system for planning of emergency evacuation. J Transport Eng 132(10):800–807

Lu Q, George B, Shekhar S (2005) Capacity constrained routing algorithms for evacuation planning: a summary of results. In: Medeiros BC, Egenhofer M, Bertino E (eds) Lecture notes in computer science, vol 3633. Springer, Berlin, pp 291–307

Mamada S, Uno T, Makino K, Fujishige S (2005) A tree partitioning problem arising from an evacuation problem in tree dynamic networks. J Oper Res Soc Jpn 48(3):196–206

Merchant DK, Nemhauser GL (1978) A model and an algorithm for the dynamic traffic assignment problems. Transport Sci 12(3):183–199

Miller-Hooks E, Patterson SS (2004) On solving quickest time problems in time-dependent, dynamic networks. J Math Model Algorithm 3:39–71

Nasrabadi E, Hashemi SM (2010) Minimum cost time-varying network flow problems. Optim Meth Software 25(3):429–447

Ng M, Waller ST (2009) The evacuation optimal network design problem: model formulation and comparison. Transport Lett: Int J Transport Res 1:111–119

Ng M, Waller ST (2010) Reliable evacuation planning via demand inflation and supply deflation. Transport Res Part E: Logist Transport Rev 46(6):1086–1094

Osman MS, Ram B, Bhadury J, Stanfield P, Davis L, Samanlioglu (2009) Optimization model for distributed routing for disaster area logistics. In: Proceedings of the 2009 IEEE international conference on service operations, logistics and informatics, Sponsored by the IEEE Intelligent Transportation Systems Society, INFORMS Service Science Section, Chicago, IL, USA, 22–29 July 2009, pp 278–283

Özdamar L, Yi W (2007) A dynamic logistics coordination model for evacuation and support in disaster response activities. Eur J Oper Res 179(3):1177–1193

Özdamar L, Yi W (2008) Greedy neighborhood search for disaster relief and evacuation logistics. IEEE Intell Syst 23(1):14–23

Pascoal MMB, Captivo MEV, Climaco JCN (2006) A comprehensive survey on the quickest path problem. Ann Oper Res 147(1):5–21

Peeta S, Kalafatas G (2008) Primary emergency routes for transportation security. Joint transportation research program, Paper 323. http://docs.lib.purdue.edu/jtrp/323

Peeta S, Kalafatas G (2009) Planning for evacuation: insights from an efficient network design model. J Infrastruct Syst 15(1):21–30

Pel AJ, Hoogendoorn SP, Bliemer MC (2010) Evacuation modeling including traveler information and compliance behavior. In: Hoogendoorn SP, Pel AJ, Taylor MAP, Mahmassani H (eds), 1st conference on evacuation modeling and management, Procedia engineering, vol 3, pp 101–111, http://www.sciencedirect.com/science/article/pii/S1877705810004807, 20.02.2011

Pérez-Villalonga F, Salmerón J, Wood K (2008) Dynamic evacuation routes for personnel on a naval ship. Nav Res Logist 55:785–799

Rebennack S, Arulselvan A, Elefteriadou L, Pardalos P (2010) Complexity analysis for maximum flow problems with arc reversals. J Combin Optim 19:200–216, 10.1007/s10878-008-9175-8

Regnier E (2008) Public evacuation decisions and hurricane track uncertainty. Manag Sci 54(1):16–28

Saadatseresht M, Mansourian A, Taleai M (2009) Evacuation planning using multiobjective evolutionary optimization approach. Eur J Oper Res 198(1):305–314

Sheffi Y (1984) Urban transportation networks: equilibrium analysis with mathematical programming techniques. Prentice Hall, NJ

Sheffi Y, Mahmassani H, Powell WB (1982) A transportation network evacuation model. Transport Res A 16(3):209–218

Sherali HD, Carter TB, Hobeika AG (1991) A location-allocation model and algorithm for evacuation planning under hurricane/flood conditions. Transport Res Part B: Methodol 25(6):439–452

Skutella M (2008) An introduction to network flows over time. In: Cook WJ, Lovasz L, Vygen J (eds) Research trends in combinatorial optimization. Springer, Berlin, pp 451–482

Southworth F (1991) Regional evacuation modeling: a state-of-the-art review. Oak Ridge National Laboratory ORNL-11740, Tennessee

Southworth F, Chin S-M (1987) Network evacuation modelling for flooding as a result of a dam failure. Environ Plann A 19:1543–1558

Stepanov A, MacGregor Smith J (2009) Multi-objective evacuation routing in transportation networks. Eur J Oper Res 198(2):435–446

Taaffe K, Kohl R, Kimbler D (2005) Hospital evacuation: issues and complexities. In: Proceedings of the 2005 winter simulation conference, ACM 2005, ISBN 0-7803-9519-0, Orlando, FL, 4–7 Dec 2005, pp 943–950

Theodoulou G, Wolshon B (2004) Alternative methods to increase the effectiveness of freeway contraflow. J Transport Res Board 1865:48–56

Tjandra SA (2003) Dynamic network flow optimization with application to the evacuation problem. Ph.D. thesis, University of Kaiserslautern

Tuydes H, Ziliaskopoulos A (2006) Tabu-based heuristic approach for optimization of network evacuation contraflow. Transport Res Rec: J Transport Res Board 1964(1):157–168

Williams BM, Tagliaferri AP, Meinhold SS, Hummer JE, Rouphail NM (2007) Simulation and analysis of freeway lane reversal for coastal hurricane evacuation. J Urban Plann Develop 133(1):61–72

Xie C, Turnquist MA (2009) Integrated evacuation network optimization and emergency vehicle assignment. Transport Res Rec: J Transport Res Board 2091:79–90

Xie C, Turnquist MA (2011) Lane-based evacuation network optimization: an integrated lagrangian relaxation and tabu search approach. Transport Res C Emerg Tech 19(1):40–63

Xie C, Lin D-Y, Waller ST (2010) A dynamic evacuation network optimization problem with lane reversal and crossing elimination strategies. Transport Res Part E: Logist Transport Rev 46(3):295–316

Yamada T (1996) A network flow approach to a city emergency evacuation planning. Int J Syst Sci 27(10):931–936

Yao T, Mandala SR, Chung BD (2009) Evacuation transportation planning under uncertainty: a robust optimization approach. Network Spatial Econ 9:171–189

Yazici MA, Ozbay K (2007) Impact of probabilistic road capacity constraints on the spatial distribution of hurricane evacuation shelter capacities. Transport Res Rec: J Transport Res Board 2022(1):55–62

Zhang H, Gao Z (2007) Two-way road network design problem with variable lanes. J Syst Sci Syst Eng 16(1):50–61

Ziliaskopoulos AK (2000) A linear programming model for the single destination system optimum dynamic traffic assignment problem. Transport Sci 34(1):37–49

Chapter 3
An Urban Evacuation Problem

In this chapter, linear mixed-integer urban evacuation models are presented. The assumption of the presented evacuation problem in this chapter is that every evacuee has the possibility to self-evacuate, i.e. all evacuees can leave the evacuation zone by vehicle. Hence a single commodity dynamic network flow model is considered. In order to capture the progress of an evacuation, discrete dynamic network flow models are used that force the evacuees to leave the evacuation area with respect to certain objectives within the considered time horizon. Because of the dynamic flow problem the time-expanded network of the detailed network is considered, i.e. every node is copied for every point in time and arcs connect the time-copies of nodes depending on the associated travel-time. For the purpose of restructuring the traffic routing we will use the following traffic management ideas: the reversal of lanes, the prohibition of crossing conflicts of lanes within intersections and the limitation of the number of merging lanes within intersections. Crossing conflicts will be prohibited and merging will be restricted within intersections in all models, but we will consider models that allow a two-way assignment of lanes and models that allow just the assignment of one-way streets.

The traffic routing is to be optimized for the case of an evacuation and is fixed over the considered time horizon. The evacuees have to leave their starting points and have to reach a destination within the time horizon. Constraints for the number of lanes that build the traffic routing have to be taken into account and constraints for flow like the flow conservation or the avoidance of irregular traffic flow will be regarded. The evacuation zone and exits to safe areas are assumed to be known.

The chapter is organized as follows: A network representation will be introduced in Sect. 3.1. A basic urban evacuation model will be presented in Sect. 3.2 (compare Bretschneider and Kimms, 2011) and extensions of the model will be provided in Sects. 3.3 and 3.4. In Sect. 3.5, examples of solutions of the different models will be discussed.

S. Bretschneider, *Mathematical Models for Evacuation Planning in Urban Areas*,
Lecture Notes in Economics and Mathematical Systems 659,
DOI 10.1007/978-3-642-28759-6_3, © Springer-Verlag Berlin Heidelberg 2013

3.1 A Network Representation

We assume that the existing street network is used for evacuation in an urban area. Some streets may already be damaged and are limited for use or are not usable, and so the network of usable streets is used as input to the planning phase. The street network consists of intersections which are connected by streets. The number of persons to be evacuated is input to the planning phase as well. These numbers need to be estimated. In Germany, for instance, the number of persons who live in a city block, the smallest area being surrounded by streets, is known.

Figure 3.1 illustrates a street network and its blocks. We assume, a simplifying assumption, that all persons escape from the dangerous area by a vehicle. Thus, let us assume that we have an estimate of the number of cars (car loads) which have to get away from a certain block.

The whole evacuation problem will be represented as a dynamic network flow problem with complicating constraints. This basically means that we will define a graph $\mathbf{G} = (\mathcal{N}, \mathcal{A})$ with sources and sinks such that flows of vehicles can be determined in this graph.

In contrast to Chap. 1, we will introduce a notation where the nodes are represented by two indices, i.e. four indices are needed to describe an arc. We decided to choose this notation to map crossing conflicts in a general way.

We will use $\mathcal{S}$ $(\mathcal{S} \subset \mathcal{N})$ to denote the set of sources and $\mathcal{D}$ $(\mathcal{D} \subset \mathcal{N})$ to denote the set of sinks. One source is located between two intersections, say, i and j. Let $O(i, j)$ be the number of vehicles which start at the corresponding source denoted by (i, j). The number of vehicles can be estimated by the number inhabitants of adjoining blocks. For notational convenience we assume that (i, j) and (j, i) represent the very same source and likewise $O(i, j) = O(j, i)$. Figure 3.1 illustrates the position of sources and sinks.

The graph $\mathbf{G} = (\mathcal{N}, \mathcal{A})$ of the street network consisting of intersections which are connected by streets is built as follows. Let $\mathcal{J}$ be the set of intersections (junctions). Every intersection $i \in \mathcal{J}$ is the value α_i of the number of entrances/exits of intersection i assigned. A typical intersection where two streets cross each other has four entrances/exits. Intersections with just one entrance/exit ($\alpha_i = 1$) represent dead end streets or egresses of the evacuation zone. Let $\mathcal{J}_\mathcal{D} \subseteq \{d \in \mathcal{J} : \alpha_d = 1\}$ be the subset of intersections that are defined to be sinks (destinations). Furthermore let $\mathcal{P} \subseteq \mathcal{J} \times \mathcal{J}$ be the set that contains all pairs of intersections that are directly connected. The elements of $\mathcal{P}$ will represent source or sink nodes and let $(i, j) \in \mathcal{P}$ and $(j, i) \in \mathcal{P}$ stand for the same node. A source is added between two neighboring intersections i and j. The pair (i, j) of the intersections is an element of $\mathcal{P}$ (i.e. $(i, j) \in \mathcal{P}$). If w.l.o.g. $\alpha_i = 1$ and $i \notin \mathcal{D}$ (i.e. intersection i is a dead end street) the source (i, j) is added to $\mathcal{S}$, but the intersection i is not considered further. (Otherwise, if $\alpha_j = 1$ and $j \notin \mathcal{D}$, then intersection j is not considered further.) A source is named by an element in $\mathcal{P}$. Hence the set $\mathcal{S} = \{(i, j) \in \mathcal{P} : i, j \notin \mathcal{J}_\mathcal{D}\}$ is the set of sources. Analogously, the set $\mathcal{D}$ of sink nodes is built: $\mathcal{D} = \{(i, j) \in \mathcal{P} : i \in \mathcal{J}_\mathcal{D} \text{ or } j \in \mathcal{J}_\mathcal{D}\} (= \mathcal{P} \setminus \mathcal{S})$.

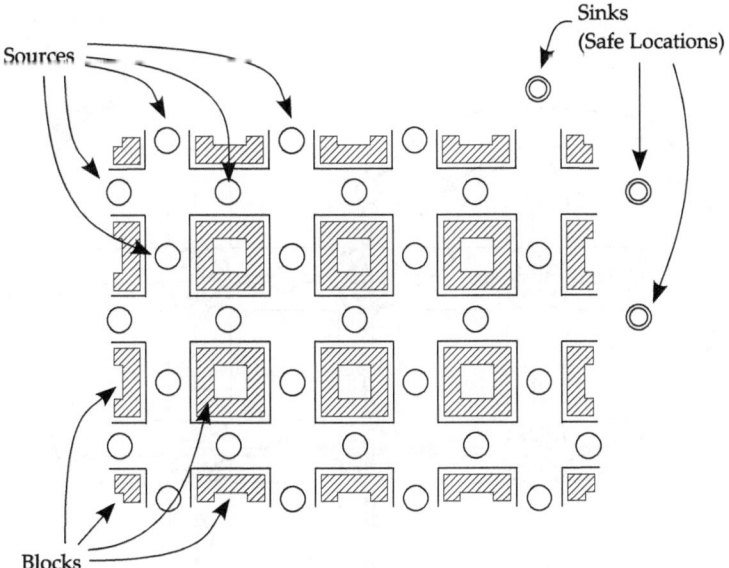

Fig. 3.1 A street network with blocks, sources and sinks—the number of inhabitants is given per block

Additionally, a super source $(S,0)$ and a set $\mathcal{A}_S$ of arcs that point from the super source $(S,0)$ at all sources in S as well as a super sink $(D,0)$ and a set $\mathcal{A}_D$ that point from a sink node $(d_1,d_2) \in \mathcal{D}$ at the super sink $(D,0)$ are added.

For the objective function minimizing the total evacuation time an artificial node and an artificial arc are added: the super–super sink $(D^*,0)$ and the arc $(D,0,D^*,0)$ leading from the super sink $(D,0)$ to the super–super sink $(D^*,0)$. With the arc $(D,0,D^*,0)$ the flow exiting the network at a certain time period is detected.

Let $\mathcal{N}^*$ be the set of super nodes. The set $\mathcal{N}^*$ is different depending on the realization of the objective functions: it is $\mathcal{N}^* = \{(S,0),(D,0)\}$ or $\mathcal{N}^* = \{(S,0),(D,0),(D^*,0)\}$. $\mathcal{A}^*$ is the set of arcs emerging from or pointing at a super node, i.e. $\mathcal{A}^* = \mathcal{A}_S \cup \mathcal{A}_D$ or $\mathcal{A}^* = \mathcal{A}_S \cup \mathcal{A}_D \cup \{(D,0,D^*,0)\}$ depending on which objective function is used. We refer to Fig. 3.2 for an illustration.

Every intersection $l \in \mathcal{J}$ with $\alpha_i > 1$ builds a sub-network constructed by α_i entrances/exits of the intersection and all arcs that point from one node to every other node of the intersection. Hence, the set of entrance/exit nodes of intersection i is $\mathcal{N}_{\mathcal{J}}(i) = \{(i,h) : h \in \{1,\dots,\alpha_i\}\}$ and the set of arcs of i is $\mathcal{A}_{\mathcal{J}}(i) = \{(i,l,i,h) \in \mathcal{N}_{\mathcal{J}}(i) \times \mathcal{N}_{\mathcal{J}}(i) : l \neq h\}$.

We assume that the numbering of the entrance/exit numbers for all intersections is clockwise. If the numbering is clockwise, then the constraints of the urban evacuation model (see Sect. 3.2) determine crossing conflicts for right-hand traffic, otherwise for left-hand traffic.

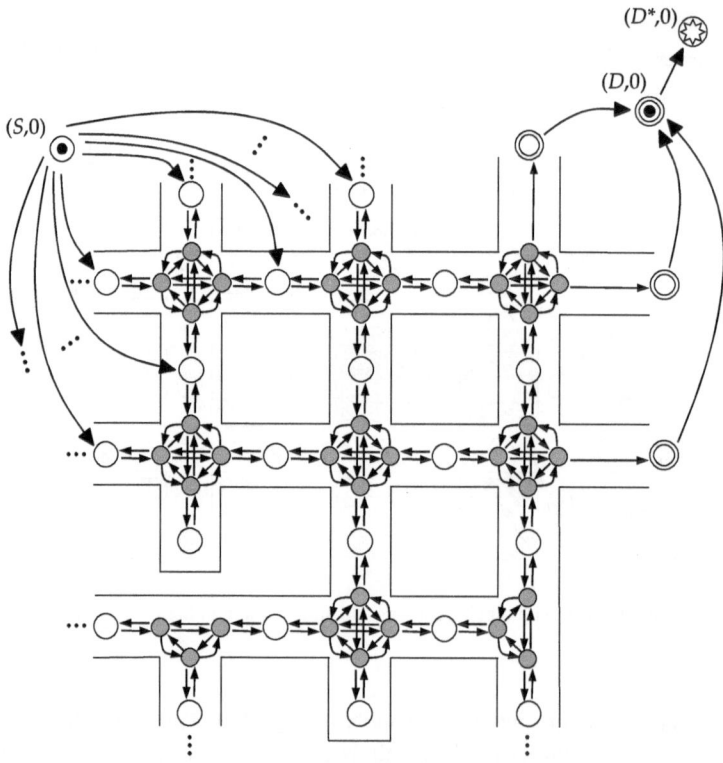

Fig. 3.2 An illustration of the nodes and the arcs in the network

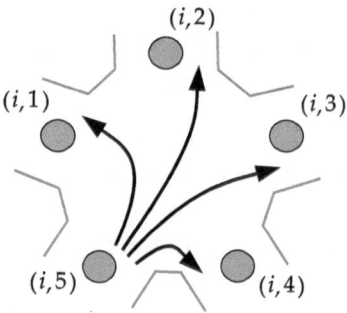

Fig. 3.3 An illustration of a left turn arc, straight forward arcs, and a right turn arc emerging from entrance $(i, 5)$ of intersection i

The set of nodes of all intersection is $\mathcal{N}_{\mathcal{J}} = \bigcup_{i \in \mathcal{J}} \mathcal{N}_{\mathcal{J}}(i)$ and the set of all arcs within intersections is the set $\mathcal{A}_{\mathcal{J}} = \bigcup_{i \in \mathcal{J}} \mathcal{A}_{\mathcal{J}}(i)$. Set $\mathcal{A}_{\mathcal{J}}^{l}$ is the set of left turn arcs within an intersection, and $\mathcal{A}_{\mathcal{J}}^{sf}$ is the set of straight forward arcs that are all arcs that do not directly turn left or right. Figure 3.3 shows the left arc $(i, 5, i, 1)$, the two straight forward arcs $(i, 5, i, 2)$ and $(i, 5, i, 3)$ and the right turn arc $(i, 5, i, 4)$ emerging from entrance 5 of intersection i.

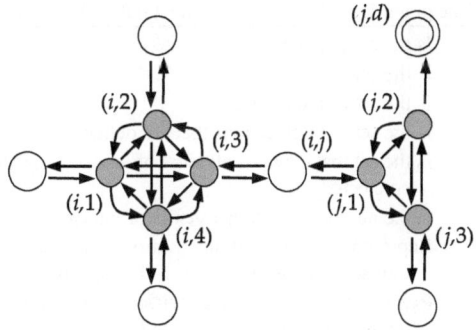

Fig. 3.4 An illustration of the arcs $\mathcal{A}_C$ and $\mathcal{A}_{\mathcal{J}}$ of the intersections i and j

Additional detailed information about the entrances/exits of an intersection is needed. $\mathcal{A}_C$ is the set of arcs that connect sources $(i, j) \in \mathcal{S}$ and sinks $(i, j) \in \mathcal{D}$ with the appropriate entrance/exit-nodes of the intersections i and j: The arcs pointing in both directions are included, if the number of entrances/exits of i and j is greater than 1. If w.l.o.g. $\alpha_i = 1$ and $i \notin \mathcal{D}$, then the arc pointing from (i, j) to the appropriate entrance-node (j, h) of the intersection j and the arc of the opposite direction are included in $\mathcal{A}_C$, i.e. (i, j, j, h) and (j, h, i, j). If $(i, j) \in \mathcal{N}_{\mathcal{D}}$ with w.l.o.g. $j \in \mathcal{D}$ then just the arc pointing from the appropriate exit-node (i, l) of intersection i to the sink node (i, j) belongs to $\mathcal{A}_C$ $((i, l, i, j) \in \mathcal{A}_C)$. (Otherwise, if $(i, j) \in \mathcal{N}_{\mathcal{D}}$ with $i \in \mathcal{D}$, then $(j, h, i, j) \in \mathcal{A}_C$, where (j, h) is an appropriate exit-node.) Figure 3.4 illustrates the notation of the nodes and arcs of a street network. Intersections i and j with 4 and 3 entrances/exits that are next to each other are depicted with the corresponding nodes $(i, 1), \ldots, (i, 4)$ and $(j, 1), (j, 2), (j, 3)$, respectively, and the 12 arcs of intersection i and the 6 arcs of intersection j. Arcs connecting these intersections are $(i, 3, i, j)$, $(i, j, i, 3)$, $(j, 1, i, j)$, and $(i, j, j, 1)$. Node (j, d) depicts a sink, where d is an "intersection" with $\alpha_d = 1$ that is determined as a sink, i.e. $d \in \mathcal{J}_{\mathcal{D}}$. Intersection j and sink d are connected by one arc, namely $(j, 2, j, d)$.

The complete street network is represented by the network $\mathbf{G} = (\mathcal{N}, \mathcal{A})$ with $\mathcal{N} = \mathcal{S} \cup \mathcal{N}_{\mathcal{J}} \cup \mathcal{D} \cup \mathcal{N}^*$ and $\mathcal{A} = \mathcal{A}_{\mathcal{J}} \cup \mathcal{A}_C \cup \mathcal{A}^*$. The notation of the components of graph $\mathbf{G}$ is listed in Table 3.1 for the sake of clarity.

3.2   A Mixed-Integer Urban Evacuation Model

3.2.1   Decision Variables and Additional Parameters

To come closer to a real-world situation, we need to add a few more aspects to the network: A certain street section, which corresponds to arcs in the set $\mathcal{A}_C \cup \mathcal{A}_{\mathcal{J}}$, has a limited number of lanes. The number of vehicles that can enter and that can

Table 3.1 Notation of the graph $\mathbf{G} = (\mathcal{N}, \mathcal{A})$ of the street network and artificial nodes and arcs

$\mathcal{N}$	Is the set of nodes
$\mathcal{A}$	Is the set of arcs
$\mathcal{J}$	Is the set of intersections
$\mathcal{J}_D \subset \mathcal{J}$	Is the set of intersections that are defined as destinations
$\mathcal{P} \subseteq \mathcal{J} \times \mathcal{J}$	Is the set that contains all pairs of intersections that are directly connected
$\alpha_i \in \mathbb{Z}^+$	Is the number of entrances/exits of intersection $i \in \mathcal{J}$
$\mathcal{N}_\mathcal{J}, \mathcal{N}_\mathcal{J}(i)$	Represent the set of all intersection nodes and the set of nodes of intersection $i \in \mathcal{J}$ with $\alpha_i > 1$, respectively
$\mathcal{A}_\mathcal{J}, \mathcal{A}_\mathcal{J}(i)$	Represent the set of all arcs within intersections and the set of arcs of intersection $i \in \mathcal{J}$ with $\alpha_i > 1$
$\mathcal{A}_\mathcal{J}^l, \mathcal{A}_\mathcal{J}^{sf}$	Represent the set of left turn arcs and straight forward arc, respectively
$\mathcal{A}_C$	Is the set of arcs that connect sources and destinations with the nodes of the intersections
$\mathcal{S}$	$= \{(i, j) \in \mathcal{P} : i, j \notin \mathcal{J}_D\}$ is the set of source nodes
$\mathcal{D}$	$= \mathcal{P} \backslash \mathcal{S}$ is the set of sink nodes
$(S, 0)$	Represents the super source node
$(D, 0), (D^*, 0)$	Represents super sink node and super–super sink node, respectively
$\mathcal{A}_S$	Is the set of arcs that point from the super source $(S, 0)$ at all sources in $\mathcal{S}$
$\mathcal{A}_D$	Is the set of arcs that point from a sink node $(d, 0) \in \mathcal{N}_D$ at the super sink $(D, 0)$
$\mathcal{N}^*$	Denotes the set of super nodes (i.e. it is either $\mathcal{N}^* = \{(S, 0), (D, 0)\}$ or $\mathcal{N}^* = \{(S, 0), (D, 0), (D^*, 0)\}$
$\mathcal{A}^*$	Is the set of arcs emerging from or pointing at a super node (i.e. it is $\mathcal{A}^* = \mathcal{A}_S \cup \mathcal{A}_D$ or $\mathcal{A}^* = \mathcal{A}_S \cup \mathcal{A}_D \cup \{(D, 0, D^*, 0)\}$)

use the street section at the same time is limited as well. In addition, vehicles need a certain time to traverse a street section. I.e. the parameters for number of lanes (no matter what direction the lanes go during normal traffic) and the travel time of each considered street segment, the inflow and total capacity of each direction of a street segment are taken into account. Let the time horizon be subdivided into $T + 1$ time units where T is an upper bound on the number of time units required to evacuate all vehicles from the dangerous zone.

Accordingly, we need the following parameters and sets to formulate the urban evacuation model:

- Let $\mathcal{T} = \{0, 1, \ldots, T\}$ be the set of equidistant time points in which snapshots of the situation are taken.
- Let $\mathcal{L}(j, l, i, h) = \mathcal{L}(i, h, j, l)$ denote the number of lanes of arcs $(j, l, i, h) \in \mathcal{A}_C \cup \mathcal{A}_\mathcal{J}$.
- Let $c(j, l, i, h)$ be the maximum number of vehicles (the total capacity) on a single lane per time unit for the street section which corresponds to arc $(j, l, i, h) \in \mathcal{A}_C \cup \mathcal{A}_\mathcal{J}$. The arcs in the set $\mathcal{A}_S \cup \mathcal{A}_D \cup \{(D, 0, D^*, 0)\}$ have unlimited capacity, because they do not represent a real street but were introduced for technical reasons (see the mathematical formulation in the following subsection, Sect. 3.2.2) only.

- Let $c^{in}(j,l,i,h)$ denote the number of vehicles (inflow capacity) which can enter a lane per time unit where $(j,l,i,j) \in \mathcal{A}_C \cup \mathcal{A}_{\mathcal{J}}$.
- Let $\tau(j,l,i,h) \in \mathbb{N}_0$ denote the estimated minimum time (number of time units) it takes to drive through the street that corresponds to an arc $(j,l,i,h) \in \mathcal{A}$. For arcs in the set $\mathcal{A}_S \cup \mathcal{A}_D$ the value τ is zero.
- $O(i,j)$ denotes the number of evacuees that are initial assigned to source $(i,j) \in \mathcal{S}$.

The evacuation problem can now be stated as a dynamic network flow problem with complicating constraints. The decision variables describing the inflow of vehicles and the traffic routing are defined as follows:

- $x(j,l,i,h,t) \in \mathbb{R}_0^+$ represents the non-negative and real-valued number of vehicles which leave node (j,l) in time period $t \in \mathcal{T}$ to reach node (i,h) in time period $t + \tau(j,l,i,h)$ with $(j,l,i,h) \in \mathcal{A}$. For the ease of notation we define $x(j,l,i,h,t) = 0$ for $t < 0$.
- $\ell(j,l,i,h) \in \mathbb{N}_0$ stands for the number of lanes to be used in a certain direction, where $(j,l,i,h) \in \mathcal{A}_C \cup \mathcal{A}_{\mathcal{J}}$.
- The binary decision variable $y(i,l,i,h) \in \{0,1\}$ indicates whether (value 1) or not (value 0) there is a positive number of lanes allowed on arc $(i,l,i,h) \in \mathcal{A}_{\mathcal{J}}^f \cup \mathcal{A}_{\mathcal{J}}^l$.

3.2.2 Model Formulation

The model can be divided into objective function, network flow, traffic routing and capacity constraints and constraints concerned with superfluous arcs.

Objective Function

We will consider the minimization of the average evacuation time of every unit of flow, i.e. every vehicle, and the minimization of the total evacuation time.

- The objective function is to minimize the weighted sum of flows on arcs to the super sink $(D,0)$ where the weights increase in time, i.e. the average evacuation time per vehicle can be determined:

$$\min \sum_{(d_1,d_2)\in\mathcal{D}} \sum_{t\in\mathcal{T}} t \cdot x(d_1,d_2,D,0,t) \tag{3.1}$$

with $\mathcal{N}^* = \{(S,0),(D,0)\}$ is the set of super nodes.

- Minimizing the total evacuation time:

$$\min \quad \Theta \tag{3.2}$$

$$s.t. \quad \Theta \geq t \cdot y_E(t) \ \forall t \in T \tag{3.3}$$

$$x(D, 0, D^*, 0, t) \leq c^{in}(D, 0, D^*, 0) y_E(t) \ \forall t \in T \tag{3.4}$$

with decision variables $\Theta \in \mathbb{R}_0^+$ and $y_E(t) \in \{0, 1\}, t \in T$ is assigned to arc $(D, 0, D^*, 0)$. $(D^*, 0)$ is the super–super sink and $c^{in}(D, 0, D^*, 0)$ is an sufficient large value. In the case of the minimization of the evacuation time, the set $\mathcal{N}^* = \{(S, 0), (D, 0), (D^*, 0)\}$ is the set of super nodes. The "price" of finding the minimum evacuation time is that there are $T + 1$ additional binary variables that indicate whether flow uses arc $(D, 0, D^*, 0)$ in a time period $t \in T$ or not.

Network Flow Constraints

- Flow enforcement:
  We need to make sure that the proper amount of flow is initiated for all sources in S and that the total amount of initial flow reaches the sinks.

$$\sum_{t \in T} x(S, 0, s_1, s_2, t) = O(s_1, s_2) \ \text{ for all } (s_1, s_2) \in S \tag{3.5}$$

$$\sum_{(d_1, d_2) \in D} \sum_{t \in T} x(d_1, d_2, D, 0, t) = \sum_{(s_1, s_2) \in S} O(s_1, s_2) \tag{3.6}$$

- Flow conservation constraints:
  Flow that reaches a node in period t has to leave it in period t, i.e. the total inflow of a node in time period t is equal to the outflow of the node in time period t.

$$\sum_{(i,h) \in \mathcal{N}:(i,h,j,l) \in \mathcal{A}} x(i, h, j, l, t - \tau(i, h, j, l)) \tag{3.7}$$

$$= \sum_{(i,h) \in \mathcal{N}:(j,l,i,h) \in \mathcal{A}} x(j, l, i, h, t)$$

$$\text{for all } (j, l) \in \mathcal{N} \backslash \widetilde{\mathcal{N}}^*; t \in T$$

where $\widetilde{\mathcal{N}}^* \subseteq \mathcal{N}^*$ is the set of super nodes that are not considered in the flow conservation constraints. Which set $\widetilde{\mathcal{N}}^*$ is chosen depends on the objective function and the corresponding graph $\mathbf{G}$ ($\widetilde{\mathcal{N}}^* := \mathcal{N}^*$ for objective function (3.1) and $\widetilde{\mathcal{N}}^* := \{(S, 0), (D^*, 0)\}$ for (3.2)).
- Avoidance of irregular flow behavior I:
  Flow within an intersection has to traverse the intersection on exactly one arc, i.e. flow within an intersection j enters through one intersection node of j and leaves

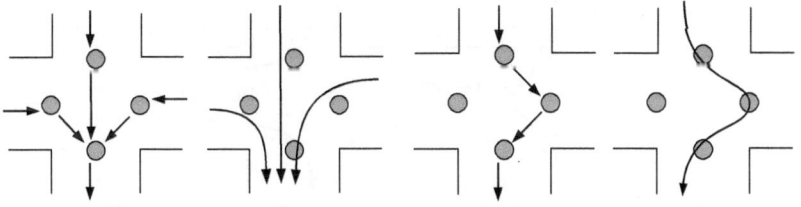

(a) Regular flow within an intersection   (b) Irregular flow within an ivntersection

Fig. 3.5 An illustration of irregular and regular flow within an intersection

through another node of j without passing any other intersection nodes of j in between. Instead of waiting earlier in the super source, flow could "wait" within intersections by using more than one intersection arc in a row (see Fig. 3.5b). Such an undesired effect may occur because of the lack of capacity. That would be like an ordered zigzag driving or turnarounds within intersections and that has to be avoided.

$$\sum_{(j,h)\in\mathcal{N}_{\mathcal{J}}\backslash\{(j,l)\}} x(j,h,j,l,t-\tau(j,h,j,l)) \le x(j,l,g,k,t) \qquad (3.8)$$

$$\text{for all } (j,l) \in \mathcal{N}_{\mathcal{J}}; (j,l,g,k) \in \mathcal{A}_C; t \in \mathcal{T}$$

Figure 3.5 illustrates a regular flow and irregular flow within an intersection. On the left side of Fig. 3.5a, b is a more technical arc-based representation, on the right side the Figures show a trajectory of a vehicle in a more intuitive way.

- Avoidance of irregular flow behavior II:
  Similar to constraint (3.8) no "waiting" or "delaying" on connecting arcs $\mathcal{A}_C$ is allowed, i.e. turnovers are prohibited within a street. No evacuee is allowed to enter a street, turn directly and drive right back into the intersection coming from to turn into another direction.

$$x(j,l,i,h,t-\tau(j,l,i,h)) \le \sum_{(g,k):(i,h,g,k)\in\mathcal{A}\backslash(i,h,j,l)} x(i,h,g,k,t) \qquad (3.9)$$

$$\text{for all } (j,l,i,h) \in \mathcal{A}_C; t \in \mathcal{T}$$

Figure 3.6 illustrates an example of an irregular flow. On the left side the traveled arcs are depicted, on the right side a trajectory of a vehicle is shown.

Traffic Routing Constraints

- Lane consistency I:
  The number of lanes on a street segment is given, but the direction in which those lanes are used is subject to our decision. The number of lanes assigned to both

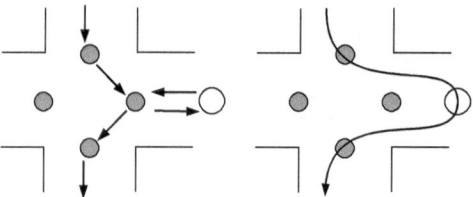

Fig. 3.6 An illustration of irregular flow on arcs in $\mathcal{A}_C$

directions of a street is bounded by the number of available lanes of the street.

$$\ell(i,l,i,h) + \ell(i,h,i,l) \le \mathcal{L}(i,l,i,h) \qquad (3.10)$$

$$\text{for all } (i,l,i,h) \in \mathcal{A}_{\mathcal{J}} : l < h$$

$$\ell(j,l,i,h) + \ell(i,h,j,l) \le \mathcal{L}(j,l,i,h) \qquad (3.11)$$

$$\text{for all } (j,l,i,h) \in \mathcal{A}_C : (i,h) \in \mathcal{S}$$

$$\ell(j,l,i,h) \le \mathcal{L}(j,l,i,h) \qquad (3.12)$$

$$\text{for all } (j,l,i,h) \in \mathcal{A}_C : (i,h) \in \mathcal{D}$$

- Lane consistency II:
  The number of lanes within an intersection that point to a certain exit of the intersection is restricted in such a way that the number of merging lanes is bounded and no crossing conflict of lanes occurs. That depends on the number of lanes emanating from that exit away from the intersection and the number of considered arcs (directions) pointing to that exit. Every possible combination of arcs pointing to the exit has to be taken into account. A merging of two lanes within an intersection is only allowed if the two lanes belong to different neighboring turning directions.

$$\sum_{(i,k,i,l)\in A^*} \ell(i,k,i,l) \le \ell(i,l,j,h) + (\delta - 1) \qquad (3.13)$$

$$\text{for all } (i,l) \in \mathcal{N}_{\mathcal{J}} ; (i,l,j,h) \in \mathcal{A}_C ; \delta \in \{1,\dots,\alpha_i - 1\};$$

$$A^* \in \{ A \subseteq \{(i,g,i,l)|g \in \{1,\dots,\alpha_i\}\} \subseteq \mathcal{A}_{\mathcal{J}} : |A| = \delta \}$$

See Fig. 3.7 for an illustration. Let every arc depicts one lane. Figure 3.7a shows a feasible assignment of the number of lanes of an intersection with $\alpha = 4$. There are three directions with a positive number of lanes pointing to the exit on the right. Therefore, two mergings are allowed, i.e. the merging of two lanes of neighboring directions. Figure 3.7b shows the three alternatives if just one lane (the dashed arcs) of the right turn direction is added. No possibility exists to add a lane without constructing a crossing conflict or a prohibited merging of lanes of the very same direction, here the right turn.

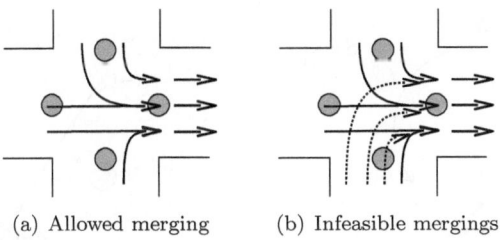

(a) Allowed merging   (b) Infeasible mergings

Fig. 3.7 An illustration of feasible and infeasible mergings

- Coupling constraints:
  If a certain direction within an intersection is allowed, i.e. the y-variable is equal to 1 (see constraints (3.15)–(3.17)), then an assignment of a positive number of lanes in this direction is possible. Right turn arcs do not lead to crossing conflicts hence we do not need to take them into account.

$$\ell(i,l,i,h) \leq \mathcal{L}(i,l,i,h)y(i,l,i,h) \text{ for all } (i,l,i,h) \in \mathcal{A}^f_{\mathcal{J}} \cup \mathcal{A}^l_{\mathcal{J}} \qquad (3.14)$$

- Prohibition of crossing conflicts I:
  No crossing of two straight forward lanes is allowed within intersections. This constraint has to be valid for all intersections with $\alpha \geq 4$.

$$y(i,l_1,i,l_2) + y(i,h_1,i,h_2) \leq 1 \quad \text{for all } (i,l_1,i,l_2) \in \mathcal{A}^{sf}_{\mathcal{J}}; \qquad (3.15)$$

$$(i,h_1,i,h_2) \in \mathcal{A}^{sf}_{\mathcal{J}} : (h_1 \in SF^{\alpha_i}_{(l_1,l_2)}, h_2 \in SF^{\alpha_j}_{(l_2,l_1)}) \text{ or}$$

$$(h_2 \in SF^{\alpha_i}_{(l_1,l_2)}, h_1 \in SF^{\alpha_i}_{(l_2,l_1)})$$

with

$$SF^{\alpha}_{(l,h)} := \begin{cases} \{k_a, \ldots, k_e\} & k_a \leq k_e \\ \{1, \ldots, \alpha\} \setminus \{k_e + 1, \ldots, k_a - 1\} & k_a > k_e \end{cases},$$

where $k_a = (l \mod \alpha)+1$ and $k_e = \alpha - ((\alpha - h + 1) \mod \alpha)$.
Figure 3.8 illustrates crossing conflicts of arc $(i,1,i,5)$ of intersection i with $\alpha_i = 6$. Each node depicts an entrance/exit of the intersection. In Fig. 3.8a the construction of the auxiliary sets $SF^6_{(1,5)}$ and $SF^6_{(5,1)}$ is depicted and Fig. 3.8b shows the resulting crossing conflicts. The black arc depicts the considered direction $(i,1,i,5)$ and the gray arcs depict the direction that build a crossing conflict with $(i,1,i,5)$.

- Prohibition of crossing conflicts II:
  No crossing of two left lanes or one left and one straight forward lane is allowed within intersections $i \in \mathcal{J}$ where $\alpha_i \geq 3$. (Intersections with $\alpha = 2$ have no crossing conflicts.)

$$y(i,l_1,i,l_2) + y(i,h_1,i,h_2) \leq 1 \qquad (3.16)$$

for all $(i,l_1,i,l_2) \in \mathcal{A}^l_{\mathcal{J}}; (i,h_1,i,h_2) \in \mathcal{A}^{sf}_{\mathcal{J}} \cup \mathcal{A}^l_{\mathcal{J}} : (h_1 = l_2) \text{ or } (h_2 = l_1)$

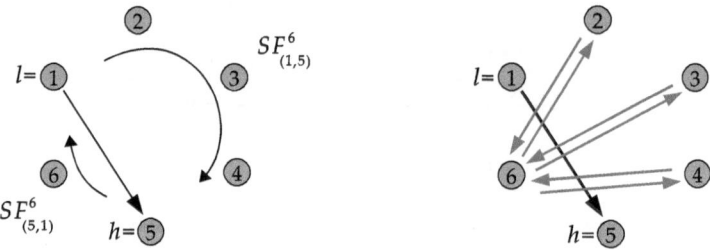

(a) Illustration of the construction of the sets $SF^6_{(1,5)}$ and $SF^6_{(5,1)}$

(b) Arcs (gray) building crossing conflicts with a straight forward arc (black)

Fig. 3.8 An illustration of crossing conflicts I of straight forward arcs of an intersection with six entrances/exits

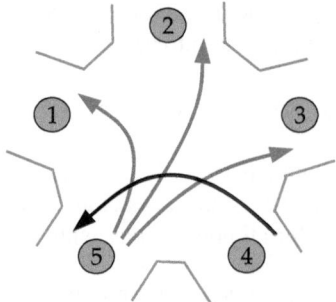

Fig. 3.9 An illustration of crossing conflicts II with a left turn arc

An example of crossing conflicts of left turn arcs with straight forward arcs or another left turn arc is given in Fig. 3.9 for intersection i with five entrances/exits: The left turn arc $(i, 4, i, 5)$ crosses with arcs $(i, 5, i, h_2) \in \mathcal{A}^{sf}_{\mathcal{J}} \cup \mathcal{A}^l_{\mathcal{J}}$ (i.e. the first case $h_1 = l_2$ of constraint (3.16) is illustrated).

- Prohibition of crossing conflicts III:
  For all intersections i with $\alpha_i \geq 5$ additional crossing conflicts of straight forward arcs that have a node in common arise and have to be prohibited:

$$y(i, l_1, i, l_2) + y(i, h_1, i, h_2) \leq 1 \tag{3.17}$$

$$(i, l_1, i, l_2) \in \mathcal{A}^{sf}_{\mathcal{J}} : l_1 \neq ((l_2 + 1) \bmod \alpha_i) + 1; (i, h_1, i, h_2) \in \mathcal{A}^{sf}_{\mathcal{J}} :$$

$$(h_1 = l_2 \text{ and } h_2 \in \{((l_2 + 1) \bmod \alpha_i) + 1, \ldots, ((\alpha_i + l_1 - 2) \bmod \alpha_i) + 1\})$$

For an example of crossing conflicts of straight forward arcs that have a node in common see the left side of Fig. 3.10 illustrating intersection i with six entrances/exits. Every node depicts an entrance/exit. The number labeling every node is the number of the entrance/exit. Arcs that built the crossing conflict described above with arc $(i, 5, i, 1)$ are the straight forward arcs

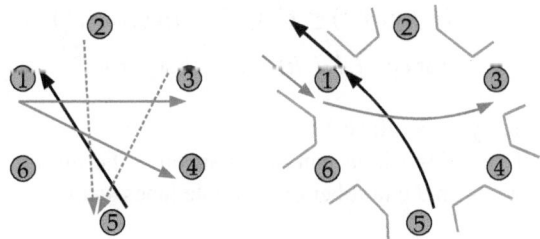

Fig. 3.10 An illustration of crossing conflicts III of straight forward arcs

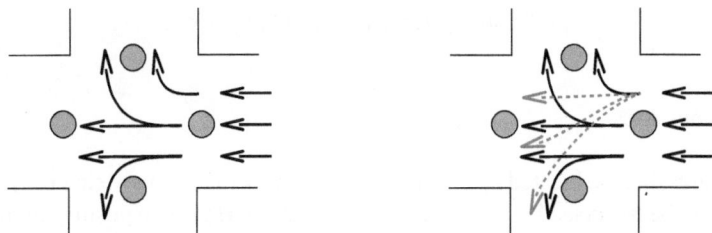

(a) Illustration of a feasible arrangement of lanes

(b) Illustration of infeasible arrangements of lanes

Fig. 3.11 An illustration of an arrangement of lanes

$(i, 1, i, 3)$, $(i, 1, i, 4)$ (depicted as gray arcs on the left in Fig. 3.10) and are determined with constraint (3.17) if $(i, l_1, i, l_2) = (i, 5, i, 1)$. The crossing conflicts $(i, 5, i, 1)/(i, 2, i, 5)$ and $(i, 5, i, 1)/(i, 3, i, 5)$ are determined if $(i, l_1, i, l_2) = (i, 2, i, 5)$ and $(i, l_1, i, l_2) = (i, 3, i, 5)$, respectively. The straight forward arcs $(i, 2, i, 5)$, $(i, 3, i, 5)$ are depicted as gray dashed arcs in Fig. 3.10. The intersection on the right of Fig. 3.10 illustrates the crossing conflict of arcs $(i, 1, i, 3)$ and $(i, 5, i, 1)$.

The crossing conflicts that we consider are crossing conflicts that are direction-based and are independent of the number of lanes. The lanes can be arranged also due to constraint "Lane consistency II" (3.13) such that no crossing conflict occurs. We assume that the vehicles order in the appropriate lanes that correspond to their subsequent turn before an intersection is entered. See Fig. 3.11a for an illustration. Figure 3.11b shows three possibilities if a lane (the gray dashed arcs) of different directions use another exit as planned. There occur crossing conflicts and hence lead to infeasible arrangements of lanes.

Capacity Constraints

- Inflow capacity constraints:
  The number of vehicles per time unit which can enter a street segment is limited and depends on the number of available lanes as well as on the inflow capacity of a single lane.

$$x(j,l,i,h,t) \leq c^{in}(j,l,i,h)\ell(j,l,i,h) \tag{3.18}$$

$$\text{for all } (j,l,i,h) \in \mathcal{A}_C \cup \mathcal{A}_{\mathcal{J}}; t \in \mathcal{T}$$

- Total street capacity constraints:
  The number of vehicles which can simultaneously be on a street segment is limited and depends on the number of available lanes again.

$$\sum_{t' \in \{t - \tau(j,l,i,h)+1,\dots,t\}} x(j,l,i,h,t') \leq c(j,l,i,h)\ell(j,l,i,h) \tag{3.19}$$

$$\text{for all } (j,l,i,h) \in \mathcal{A}_C \cup \mathcal{A}_{\mathcal{J}}; t \in \mathcal{T}$$

Superfluous Arcs

Certain arcs that are located next to an intersection i with just one exit, i.e. $\alpha_i = 1$ do not have to be traversed by flow in a feasible solution. Hence no positive numbers of lanes need to be assigned to such arcs, i.e. the number of lanes ℓ belonging to these arcs can be set to zero. Intersections with just one exit are dead ends that describe blind alleys or the end of the observed danger zone with no entrance to a safe zone. There are three cases.

- Superfluous arcs I:
  Arcs emerging from an intersection and pointing to a dead end within the evacuation area, i.e. pointing to a source $(i, j) \in \mathcal{S}$ with $\alpha_i = 1$ or $\alpha_j = 1$ are not traveled by flow in a feasible solution in any point in time due to the constraints "Avoidance of irregular flow behavior I" (3.8). We set the number of lanes corresponding to these arcs to zero, see Fig. 3.12a for an illustration (the black dashed arc).

$$\ell(j,l,i,h) = 0 \tag{3.20}$$

$$\text{for all } (j,l,i,h) \in \mathcal{A}_C : ((i,h) \in \mathcal{S} \text{ and } (\alpha_i = 1 \text{ or } \alpha_h = 1))$$

- Superfluous arcs II:
  Intersection arcs that point to an exit leading to a dead end within the evacuation area are also not traversed by flow in a feasible solution due to the constraints "Avoidance of irregular flow behavior I + II" ((3.8) and (3.9)). The number of lanes ℓ of these arcs can be set to zero. We refer to Fig. 3.12a for an illustration (black arcs).

$$\ell(i,l,i,h) = 0 \tag{3.21}$$

for all $(i,l,i,h) \in \mathcal{A}_{\mathcal{J}}$ for which $(i,h,i,j) \in \mathcal{A}_C$ with $(i,j) \in \mathcal{S}$ and $\alpha_j = 1$ exists

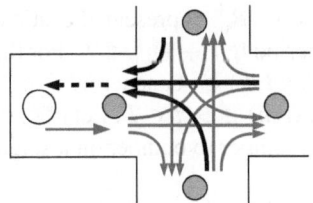 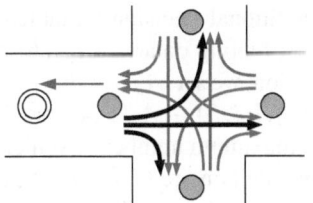

(a) Illustration of superfluous arcs I and II (dashed and solid black arcs)

(b) Illustration of superfluous arcs III (black arcs)

Fig. 3.12 Illustration of superfluous arcs

- Superfluous arcs III:
  Arcs within an intersection emerging from an exit that are located directly next to a sink are not needed, see Fig. 3.12b for an illustration (the black arcs). The numbers of lanes corresponding to these intersection arcs can be set to zero.

$$\ell(i,l,i,h) = 0 \tag{3.22}$$

for all $(i,l,i,h) \in \mathcal{A}_{\mathcal{J}}$ for which $(i,l,i,j) \in \mathcal{A}_C$ with $(i,j) \in \mathcal{D}$ exists

The solution of this evacuation model may include a positive number of lanes although in no point in time is flow assigned to the corresponding arc. These lanes do not harm the solution in terms of the objective function, but they are not needed to let flow traverse the arcs in the network optimally. One possibility is to solve the basic evacuation problem and then set all ℓ-variables of lanes belonging to an arc that is not traversed by flow in any point in time equal to zero in a simple postprocessing.

3.3   An Extension of the Urban Evacuation Model: Waiting on Arcs

In the urban evacuation model no waiting except waiting in the super source is allowed. In Sect. 1.2 a network flow model with waiting on arcs is presented. A certain amount of flow has the possibility to wait after traveling a minimum amount of time "at the end" of an arc. In other words, if flow enters the head node after the travel time τ, then there is the possibility to wait there in boxes that are only reserved for flow coming from the corresponding arc.

The dynamic network flow model described in Sect. 1.2 builds the basis of the urban evacuation model with waiting. Constraints concerning the traffic routing, the capacity constraints of the inflow and superfluous arcs remain the same as in the urban evacuation model without waiting. Capacity constraints concerning the waiting have to be included and the network flow constraints have to be adjusted.

The additional decision variables $w(j,l,i,h,t) \in \mathbb{R}_0^+$ represent the number of vehicles waiting on arc $(j,l,i,h) \in \mathcal{A}$ in time period $[t, t+1)$, $t \in \mathcal{T}$. For the ease of notation, we set $w(j,l,i,h,-1) = 0$ for all $(j,l,i,h) \in \mathcal{A}$ and $w(j,l,i,h,t) = 0$ for all $(j,l,i,h) \in \mathcal{A}_S \cup \mathcal{A}_D$ or $(j,l,i,h) \in \mathcal{A}: (i,h) \in \mathcal{D}, t \in \mathcal{T}$. Let $c^w(j,l,i,h)$ be the maximum number of vehicles that can be bumper to bumper on a single lane per time unit on arc $(j,l,i,h) \in \mathcal{A}_C \cup \mathcal{A}_J$.

Including the possibility to wait arc-based, the urban evacuation model is modeled as following.

Objective Function

The objective is the minimization of the average evacuation time per evacuee (3.1) or the minimization of the total evacuation time (3.2).

Network Flow Constraints

- Flow conservation constraints:
  Flow that reaches node (j,l) in period t or waits at the end of an arc pointing to (j,l) has to leave node (j,l) in period t or has to wait on the corresponding arc pointing to (j,l) for another period.

$$\sum_{(i,h)\in\mathcal{N}:(i,h,j,l)\in\mathcal{A}} (x(i,h,j,l,t-\tau(i,h,j,l)) + w(i,h,j,l,t-1)) \quad (3.23)$$

$$= \sum_{(i,h)\in\mathcal{N}:(j,l,i,h)\in\mathcal{A}} x(j,l,i,h,t) + \sum_{(i,h)\in\mathcal{N}:(i,h,j,l)\in\mathcal{A}} w(i,h,j,l,t)$$

$$\text{for all } (j,l) \in \mathcal{N}\backslash\widetilde{\mathcal{N}}^*; t \in \mathcal{T}$$

where $\widetilde{\mathcal{N}}^*$ is the appropriate set of super nodes depending on the chosen objective function and the corresponding graph $\mathbf{G}$ ($\widetilde{\mathcal{N}}^* := \mathcal{N}^*$ for objective function (3.1) and $\widetilde{\mathcal{N}}^* := \{(S,0),(D^*,0)\}$ for (3.2)).

- Waiting on the appropriate arc:
  Flow that waits in t on an arc $(j,l,i,h) \in \mathcal{A}_C \cup \mathcal{A}_J$ has to be waiting there already or reaches the end of arc (j,l,i,h) in t, i.e. the amount of flow waiting on (j,l,i,h) must not exceed the amount of flow waiting on (j,l,i,h) in time $t-1$ plus the amount of flow reaching the end of arc (j,l,i,h). This condition avoids that flow "jumps" to other waiting boxes than the corresponding own box.

$$w(j,l,i,h,t) \le w(j,l,i,h,t-1) + x(j,l,i,h,t-\tau(j,l,i,h)) \quad (3.24)$$

$$\text{for all } (j,l,i,h) \in \mathcal{A}_C \cup \mathcal{A}_J; t \in \mathcal{T}$$

- Avoidance of irregular flow behavior I:
  Flow that comes from an intersection arc (i, l, i, h) has to wait there or leave the intersection through the appropriate exit (i, h).

$$\sum_{(i,l,i,h)\in\mathcal{A}_\mathcal{J}} (x(i,l,i,h,t - \tau(i,l,i,h)) + w(i,l,i,h,t - 1)) \qquad (3.25)$$

$$= x(i,h,g,k,t) + \sum_{(i,l,i,h)\in\mathcal{A}_\mathcal{J}} w(i,l,i,h,t)$$

$$\text{for all } (i,h) \in \mathcal{N}_\mathcal{J} : \alpha_i > 1; (i,h,g,k) \in \mathcal{A}_C; t \in \mathcal{T}$$

- Avoidance of irregular flow behavior II:
  Flow that comes from an arc $(j, l, i, h) \in \mathcal{A}_C$ connecting two intersections or waits on this arc can start to travel through an incident arc except its counter arc (the counter arc is (i, h, j, l)) or can wait on $(j, l, i, h) \in \mathcal{A}_C$ for another time period.

$$x(j,l,i,h,t - \tau(j,l,i,h)) + w(j,l,i,h,t - 1) \qquad (3.26)$$

$$\leq \sum_{(g,k)\in\mathcal{N}:(i,h,g,k)\in\mathcal{A}_C\cup\mathcal{A}_\mathcal{J}\backslash\{(i,h,j,l)\}} x(i,h,g,k,t) + w(j,l,i,h,t)$$

$$\text{for all } (j,l,i,h) \in \mathcal{A}_C; t \in \mathcal{T}$$

Traffic Routing Constraints

The constraints "Lane consistency I" (3.10)–(3.12) that restrict the assigned number of lanes to the available number of lanes, the constraints "Lane consistency II" (3.13) that restrict the number of lanes within intersection depending on considered directions and the constraints "Coupling constraints" (3.14) as well as "Prohibition of crossing conflicts I, II, III" (3.15)–(3.17) build the traffic routing constraints.

Capacity Constraints

Beside the inflow and total traffic flow constraints (3.18) and (3.19) the following constraint is considered.

- Total waiting capacity:
  This constraint restricts the total number of vehicles on an arc, which are the vehicles traveling on the arc and vehicles waiting on the arc.

$$\sum_{t'=t-\tau(j,l,i,h)+1}^{t} x(j,l,i,h,t') + w(j,l,i,h,t) \leq c^w(j,l,i,h) \cdot \ell(j,l,i,h)$$

$$(3.27)$$

$$\text{for all } (j,l,i,h) \in \mathcal{A}_C \cup \mathcal{A}_{\mathcal{J}}; t \in \mathcal{T}$$

Superfluous Arcs

The number of lanes of superfluous arcs are set to zero are the same as for the model without waiting (these are constraints (3.20)–(3.22)).

3.4 An Extension of the Urban Evacuation Model: A One-Way Evacuation Model

An additional constraint could be that every street is readjusted to a one-way street, whereas a street can be divided into two parts and the parts can lead to the opposite directions. In the urban evacuation model two-way streets are allowed but counter traffic may cause confusion and may irritate the evacuees.

Just small changes need to be done to extend the urban evacuation model into a one-way urban evacuation model. The constraints "Prohibition of crossing conflicts II + III" (3.16) and (3.17) are no longer necessary. Such crossing conflicts only occur if counter flow is allowed and an intersection node is an entrance and an exit in the same time. Compare the modeling of the constraints (3.16) and (3.17) and see Figs. 3.9 and 3.10 for illustration.

The additional traffic management idea of readjusting all streets into one-way streets leads to some changes in the presented urban evacuation model and the variables can be adjusted as follows:

- Additional binary variables for the arcs connecting intersections are introduced: $y_C(j,l,i,h)$ for all $(j,l,i,h) \in \mathcal{A}_C$
- The binary variables indicating whether a positive number of lanes is allowed to be assigned to an intersection arc have to be defined also for right turn arcs: $y(i,l,i,h)$ for all $(i,l,i,h) \in \mathcal{A}_{\mathcal{J}}$
- The integer variables ℓ just need to be defined for all arcs within an intersection: $\ell(i,l,i,h) \in \mathbb{N}_0$ for all $(i,l,i,h) \in \mathcal{A}_{\mathcal{J}}$

Then the one-way evacuation problem can be modeled as follows. The network flow constraints for the one-way model can stay the same as in the two-way case presented in Sect. 3.2, but the restrictions on the traffic routing have to be adjusted.

Objective Functions

The average evacuation time per evacuee (3.1) or the total evacuation time (3.2) is minimized.

Network Flow Constraints

The flow enforcement constraints (3.5) and (3.6), the flow conservation constraints (3.7) and the avoidance of irregular flow behavior I and II (3.8) and (3.9) have to be regarded.

Traffic Routing Constraints

- Coupling constraints:

$$\ell(i,l,i,h) \leq \mathcal{L}(i,l,i,h)y(i,l,i,h) \text{ for all } (i,l,i,h) \in \mathcal{A}_{\mathcal{J}} \qquad (3.28)$$

$$\ell(j,l,i,h) \leq \mathcal{L}(j,l,i,h)y_C(j,l,i,h) \text{ for all } (j,l,i,h) \in \mathcal{A}_C \qquad (3.29)$$

- Lane consistency I:
  The constraint restricting the number of lanes of arcs within intersection (3.10) and (3.12) are no longer needed and the following constraints allowing just one-way streets are added.

$$y(i,l,i,h) + y(i,h,i,l) \leq 1 \text{ for all } (i,l,i,h) \in \mathcal{A}_{\mathcal{J}} : l < h \quad (3.30)$$

$$y_C(j,l,i,h) + y_C(i,h,j,l) \leq 1 \text{ for all } (j,l,i,h) \in \mathcal{A}_C : (i,h) \in \mathcal{S} \quad (3.31)$$

- Lane consistency II changes to

$$\sum_{(i,k,i,l)\in A^*} \ell(i,k,i,l) \leq \mathcal{L}(i,l,j,h)y_C(i,l,j,h) + (\delta - 1) \qquad (3.32)$$

$$\text{for all } (i,l) \in \mathcal{N}_{\mathcal{J}}; (i,l,j,h) \in \mathcal{A}_C; \delta \in \{1,\dots,\alpha_i - 1\};$$

$$A^* \in \{A \subseteq \{(i,g,i,l)|g \in \{1,\dots,\alpha_i\}\} \subseteq \mathcal{A}_{\mathcal{J}} : |A| = \delta\}$$

- The constraints "Prohibition of crossing conflicts I" (3.15) have to be regarded.

Capacity Constraints

Either the available number of lanes or no lane is assigned to one direction of every street segment that connects neighboring intersections.

- Inflow capacity constraints (3.18) for arcs in $\mathcal{A}_{\mathcal{J}}$ stay the same, but for arcs in $\mathcal{A}_C$ they change to

$$x(j,l,i,h,t) \le c^{in}(j,l,i,h)\mathcal{L}(j,l,i,h)y_C(j,l,i,h) \qquad (3.33)$$

$$\text{for all } (j,l,i,h) \in \mathcal{A}_C; t \in \mathcal{T}$$

- Total capacity constraints (3.19) for arcs in $\mathcal{A}_{\mathcal{J}}$ stay the same, but for arcs in $\mathcal{A}_C$ they change to

$$\sum_{\tilde{t} \in \{t-\tau(j,l,i,h)+1,\dots,t\}} x(j,l,i,h,\tilde{t}) \le c(j,l,i,h)\mathcal{L}(j,l,i,h)y_C(j,l,i,h) \qquad (3.34)$$

$$\text{for all } (j,l,i,h) \in \mathcal{A}_C; t \in \mathcal{T}$$

Superfluous Arcs

Constraints setting the number of lanes of superfluous arcs to zero (i.e. the constraints (3.20)–(3.22)) have to be regarded.

Another Model of the One-Way Urban Evacuation Problem

An easier realization of the idea that all streets of the readjusted traffic routing are one-way streets is described in the following. The variables y_C ($y_C(j,l,i,h)$ for all $(j,l,i,h) \in \mathcal{A}_C$) are added, the domain of definition of the variable y is extended to all arcs $\mathcal{A}_{\mathcal{J}}$ ($y(i,l,i,h)$ for all $(i,l,i,h) \in \mathcal{A}_{\mathcal{J}}$) and the integer variables ℓ of number of lanes remain as they are defined in the urban evacuation model presented in Sect. 3.2. Then just the following constraints have to be added to the urban evacuation model introduced in Sect. 3.2:

- Coupling constraints for all arcs $\mathcal{A}_{\mathcal{J}} \cup \mathcal{A}_C$:

$$\ell(i,l,i,h) \le \mathcal{L}(i,l,i,h)y(i,l,i,h) \text{ for all } (i,l,i,h) \in \mathcal{A}_{\mathcal{J}} \qquad (3.35)$$

$$\ell(j,l,i,h) \le \mathcal{L}(j,l,i,h)y_C(j,l,i,h) \text{ for all } (j,l,i,h) \in \mathcal{A}_C \qquad (3.36)$$

- Only one-way streets are allowed:

$$y(i,l,i,h) + y(i,h,i,l) \le 1 \text{ for all } (i,l,i,h) \in \mathcal{A}_{\mathcal{J}} : l < h \qquad (3.37)$$

$$y_C(j,l,i,h) + y_C(i,h,j,l) \le 1 \text{ for all } (j,l,i,h) \in \mathcal{A}_C : (i,h) \in \mathcal{S} \qquad (3.38)$$

We used that realization in the computational study in Sect. 4.3.

3.5 Examples of the Variations of the Urban Evacuation Model

In this section examples of solutions of the introduced evacuation models are presented and compared. The solutions of the following variations of the urban evacuation model are compared:

- The urban evacuation models without waiting with objective (3.1) and with objective (3.2), i.e. the minimization of the weighted sums and the minimization of the evacuation time.
- The models based on the dynamic network flow model with and without waiting with the objective function (3.1).
- The models without waiting with one-way constraints and the model without waiting with the permission to use both directions with the objective function (3.1).

The "triple optimization result" (see Jarvis and Ratliff 1982, and compare approaches of the result for several sources e.g. Chalmet et al. 1982, Hamacher and Tufekci 1987, Baumann and Skutella 2006) states that if the total amount of flow units assigned to the sources can reach the sink within the given time horizon, then the solution which minimizes the weighted sum (where the weights are increasing with time) of flows entering the sinks also minimizes the network clearance time. The flow conservation and inflow capacity constraints have to be regarded. I.e. the evacuation time is minimized in the same time as the weighted sum of flows (but not vice versa). This result is not valid for the urban evacuation model without waiting, i.e. the minimum dynamic cost flow with the above constraints is extended with the assignment of the capacities, the total capacity on arcs and traffic flow constraints. Optimizing the urban evacuation model, a subnetwork with restrictive rules on the network has to be determined such that the chosen objective function is minimized. In Sect. 3.5.1 an example is given where the minimum evacuation time is not achieved with the minimization of the weighted sum of flows (minimizing the total sum of times every evacuee is within the street network).

Fleischer and Skutella (2002) proof that a minimum dynamic cost model with the possibility of waiting in nodes does not need waiting. But for our evacuation model with waiting on arcs, examples exist where the optimal value of the function (3.1) of the model including waiting is better than without waiting. In Sect. 3.5.2 an example will be given. In Sect. 4.3 a computational study will be presented where relaxations of the urban evacuation model with and without waiting are solved. The flow models underlying the both evacuation models remain intact in this relaxation. There are 90 considered instances, where in 74.44% of the cases the objective function values are the same of the models with and without waiting and in 25.56% of the cases the objective function values of the relaxation with waiting are slightly better than without waiting. Values that differ from each other after eight decimal place are considered to be equal (there are five of these cases, in one of these five cases the value of the objective without waiting is better than the value of the objective with

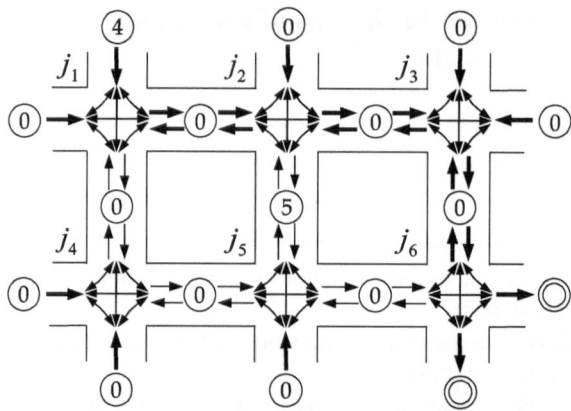

Fig. 3.13 Illustration of the network of the example comparing different objectives

waiting). The average percentage gap of the instances where the objective values are different amounts to 0.03%.

For the model allowing two-way streets and the model with one-way streets, it is expected that the value of the objective function of the two-way model is better, i.e. lesser, than the value of the objective function of the one-way model. Such an example will be illustrated in Sect. 3.5.3.

All models are implemented with AMPL using CPLEX 10.0. A computer running Windows XP with an AMD Athlon(tm) 64X2 Dual Core Processor 4600 + (2.41 GHz), 1.96 GB RAM was used for the computation.

3.5.1 An Example Considering Different Objective Functions

The urban evacuation model is based on a dynamic network flow model. The total capacity, the avoidance of irregular traffic flow and the traffic routing constraints are added. For the dynamic network flow model the "triple optimization result" is valid. That also means if the weighted sum of flows leading to the super sink is minimized then in the same time the network clearance time (total evacuation time). For the urban evacuation problem where the capacities are not assigned before optimization, it is no longer valid. An example is presented that solves the minimization of the weighted sum (3.1) optimally, but the resulting total evacuation time is not the minimum total evacuation time.

Figure 3.13 illustrates the network of the example: every arc depicts a direction, both directions of a street section within the intersections are depicted with a double arrow. The sources are mapped as nodes and the sinks that represent safe zones are mapped as nodes with double circles. The values within the sources depict the initial number of vehicles. Intersections with $\alpha = 4$ are denoted by $j_1, j_2, \ldots, j_6$. The parameters were chosen as follows:

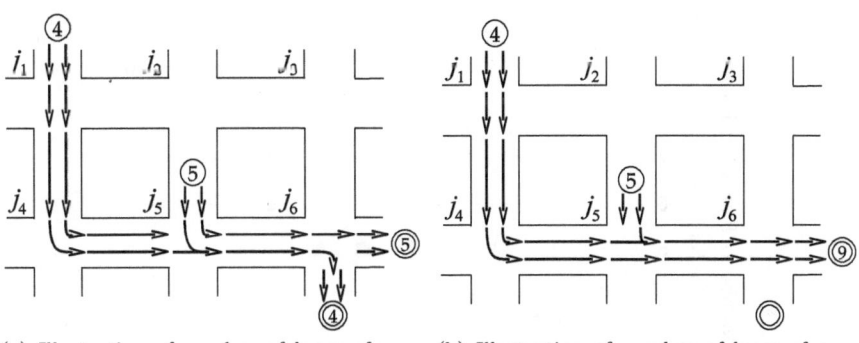

(a) Illustration of number of lanes of an optimal solution with objective (3.1)

(b) Illustration of number of lanes of an optimal solution with objective (3.2)

Fig. 3.14 Solutions of numbers of lanes of different objective functions

- There are two lanes available on every street segment.
- The inflow c^{in} per lane is set to one for every arc.
- The total flow capacity c is two and the travel time τ is three for the bold arcs, for all other arcs including the arcs within the intersections the total capacity c and the travel time τ are equal to one (see Fig. 3.13).
- The time horizon is divided into $T = 40$ periods of equal length.
- The initial numbers of evacuees are depicted in Fig. 3.13.

The objective function value of the urban evacuation problem with objective function (3.1) is 114 and the evacuation time 19 time units. Let the evacuation time of a solution be the point in time the last flow unit enters a sink. The objective function value of the urban evacuation problem with objective function (3.2) minimizing the total evacuation time is 17 time units. The value of the objective function (3.1) of the optimal solution of the urban evacuation problem minimizing (3.2) is 126.

In Fig. 3.14 only the number of lanes of street segments are drawn which are traveled by flow. The values within the circles representing the sinks denote the number of vehicles using the corresponding exit of the evacuation zone. The only significant differences of the solutions are the numbers of lanes within intersection j_5. The numbers of lanes of the solution corresponding to (3.1) give the flow from above more capacity (just one straight forward lane and two left-turn lanes), and the numbers of lanes of the solution corresponding to objective (3.2) assigns more capacity (two straight forward lanes and just one left-turn lane) to the straight forward direction emanating from the left.

Fixing the number of lanes and solving the optimization problems with different objective functions yields to the following results: If the traffic routing of the solution corresponding to objective (3.1) is fixed then the optimal solution value of the urban evacuation model with objective (3.1) is again 114 with evacuation time 19 and the optimal solution with the objective (3.2) yields to the evacuation time 19 and the corresponding value of function (3.1) is 153. If the resulting traffic

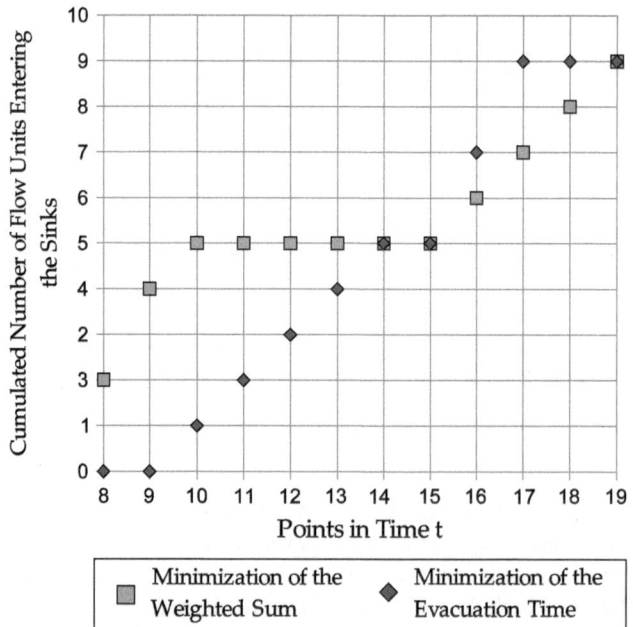

Fig. 3.15 Arrival pattern of number of evacuees leaving the evacuation zone with different objective functions

routing of the optimal solution corresponding to objective (3.2) is fixed then the solution with (3.2) yields to an evacuation time of 17 and the value of function (3.1) is 126, solving the urban evacuation problem with objective (3.1) and the fixed traffic routing, the objective value is 116 with an evacuation time of 17 time units.

The arrival pattern, the cumulated numbers of flows entering the sinks until point in time t, $t \in T$, are illustrated in Fig. 3.15. The first point in time where flow units leave the evacuation network in the optimal solution corresponding to objective (3.1) is $t = 8$. Until point in time $t = 14$ there are more evacuees in the sinks with objective (3.1), but in point in time $t = 16$ the total number of evacuees in the sinks corresponding to the solution of (3.2) is greater. The network clearing ends in point in time $t = 17$ in the solution of objective (3.2) and in $t = 19$ in the solution of objective (3.1).

3.5.2 An Example Considering Evacuation Models with and without Waiting

The network considered is the same as in the subsection before. The initial number of evacuees and the available number of lanes differ.

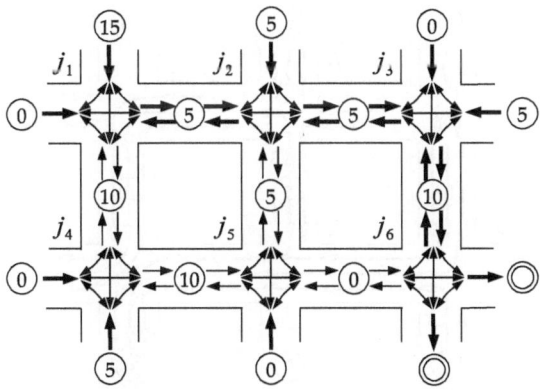

Fig. 3.16 Illustration of the network and initial number of evacuees

The parameters were chosen as follows:

- There are three lanes available on every street segment.
- The inflow c^{in} per lane is one for every arc.
- The total flow capacity c is two, the travel time τ is three and the waiting capacity c^w is seven for the bold arcs, for all other arcs including the arcs within the intersections the total capacity c and the travel time τ are equal to one and the waiting capacity c^w is three (see Fig. 3.16).
- The time horizon is divided into $T = 40$ periods of equal length.
- The initial numbers of evacuees are depicted in Fig. 3.16.

The results of the urban evacuation model with different underlying network flow models, i.e. without waiting and with waiting on arcs, are considered. The solutions of the traffic routing, i.e. the values of number of lanes of the optimal solution, respectively, are nearly the same. See Fig. 3.17 for an illustration. Each arc depicts a lane, the dashed arcs are just used by flow in the optimal solution of the urban evacuation model without waiting. Only the lanes on the directions that are used by flow over the time horizon are depicted. The names $j_1, j_2, \ldots, j_6$ depict the names of the intersection.

Some interesting results are the following: Within intersection j_1 just two lanes are assigned even if the preceding and succeeding number of lanes are three. The three lanes are not needed for the flow in this model (i.e. the corresponding capacities). The inflow rate of the left-turn in the subsequent intersection j_4 is with two number of lanes just two. The number of lanes does not need to be minimal in the model such that the number of lanes leading to and emerging from j_1 can be three. It is possible to set the number of lanes within intersection j_1 to three. This would lead to another optimal solution (the capacity is not decreased, no cost arise using different number of lanes). Merging and diverging of lanes shall be allowed, but the assignment of alternating number of lanes, if not necessary, is undesirable. Intersection j_6 does seem to have too much number of lanes, here, properly, barriers or traffic signs have to make sure that the traffic routing stays clear.

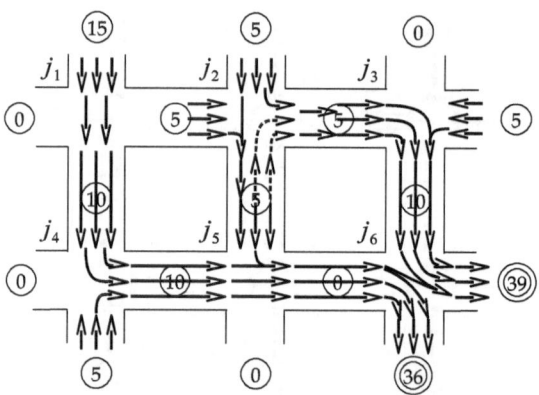

Fig. 3.17 Illustration of the number of lanes of an optimal solution

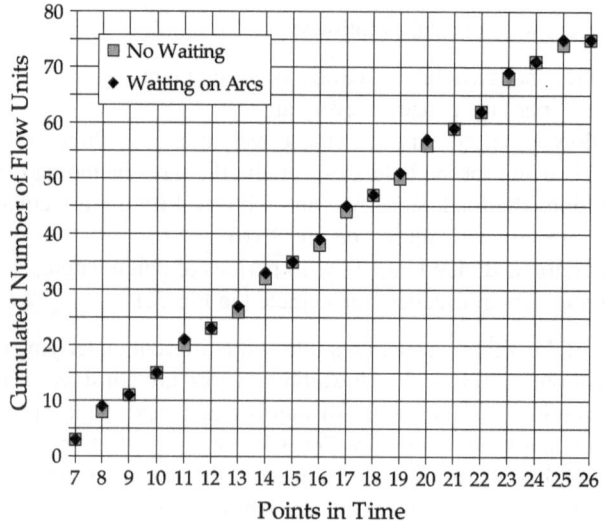

Fig. 3.18 Arrival patterns of number of evacuees leaving the evacuation zone of solutions without and with waiting

The objective function value for the optimal solution of the model without waiting is 1, 208 with an evacuation time of 26 time units. The optimal solution value of the model with waiting is 1, 197 with an evacuation time of 25 time units. See Fig. 3.18 for an illustration of the total sum of vehicles entering the sinks up to time t, $t \in \mathcal{T}$. The flows of the solution of the model with waiting are in points in time 8, 11, 13, 14, 16, 17, 19, 20, 23, 25 one vehicle ahead than the flows of the solution of the model without waiting.

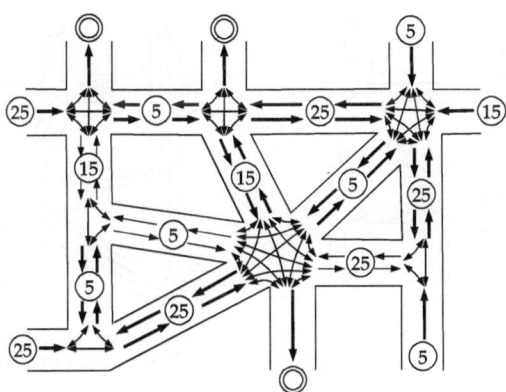

Fig. 3.19 Illustration of the network and initial number of evacuees

3.5.3 An Example Considering the Two-Way and One-Way Model

An example of a network with different numbers of exits and entrances of intersections is considered to illustrate solutions of the one-way and two-way model without waiting with respect to the objective function (3.1). The considered street network is depicted in Fig. 3.19 with the initial number of vehicles in the sources. A double circle depicts a sink. The parameters were chosen as follows:

- There are three lanes available on every street segment.
- The inflow c^{in} per lane is one for every arc.
- The total flow capacity c is five, the travel time τ is seven for the bold arcs, c is three and τ is four for the thin arcs and for the arcs within the intersections the total capacity c and travel time τ are equal to one (see Fig. 3.19).
- The time horizon is divided into $T = 50$ periods of equal length.
- The initial numbers of evacuees are depicted in Fig. 3.19.

The value of the objective function of the considered instance of the two-way model is 6, 145 with an evacuation time of 45 time units, and the value of the objective of the one-way model is 6, 152 with an evacuation time of 46 time units. The outcome of the two-way model leads to a slightly "better" solution in terms of the objective function and the evacuation time.

The solution of the two-way model is depicted in Fig. 3.20a and the solution of the one-way evacuation model is depicted in Fig. 3.20b. The resulting traffic routing in the intersection with $\alpha = 6$ entrances of the model that allows two-way streets may be a little bit confusing (see Fig. 3.20a). The traffic routing has to be made clear for examples with barricades.

In the considered instances there are few instances where the objective value of the one-way and the two-way model differ. We have considered a computational

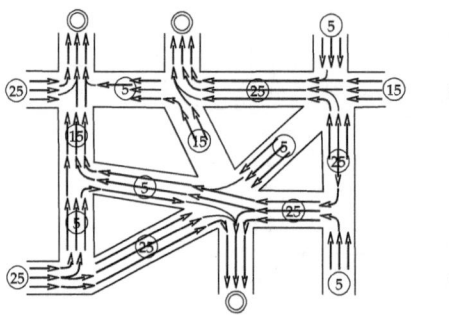
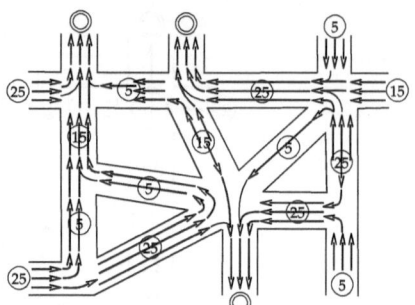

(a) Illustration of number of lanes of an   (b) Illustration of number of lanes of an
optimal solution of the two−way model   optimal Solution of the one−way model

Fig. 3.20   Solutions of numbers of lanes of the two- and one-way evacuation model

study with 60 instances. The instances are based on the above presented instance.
Additionally, the number of available lanes $\mathcal{L}$ are varied, i.e. they are for all arcs
k with $k = 2, 3, 4$, and ten different number of evacuee patterns are considered
that are chosen randomly from the set $\{5, 10, 15, 20, 25\}$. The capacity and the
travel time parameters are chosen as described above (as a first capacity/travel time
pattern) and they are changed as follows (for a second capacity/travel time pattern):
the total capacity c is equal to five for all arcs in set $\mathcal{A}_C$ and τ is equal to seven
for all arcs in set $\mathcal{A}_C$. In 90% of the optimal solutions the objectives are equal and
in 10% of the solutions the objective value of the one-way model is better than
the objective value of the two-way model. On average the percentage gap of these
objectives values with respect to the objective value of the two-way model are 0.1%.

References

Baumann N, Skutella M (2006) Solving evacuation problems efficiently–earliest arrival flows
    with multiple sources. In: 47th annual IEEE symposium on foundations of computer sci-
    ence (FOCS'06) Proceedings. IEEE Computer Society 2006, ISBN 0-7695-2720-5, Berkeley,
    California, 21–24 Oct 2006, pp 399–410
Bretschneider S, Kimms A (2011) A basic mathematical model for evacuation problems in urban
    areas. Transport Res A Pol Pract 45:523–539
Chalmet LG, Francis RL, Saunders PB (1982) Network models for building evacuation. Manag
    Sci 28(1):86–105
Fleischer L, Skutella M (2002) Minimum cost flows over time without intermediate storage.
    In: Proceedings of the 14th annual ACM-SIAM symposium on discrete algorithms, Pub-
    lisher Society for Industrial and Applied Mathematics Philadelphia, PA, ACM/SIAM 2003,
    ISBN 0-89871-538-5, 12–14 Jan 2003, pp 66–75
Hamacher HW, Tufekci S (1987) On the use of lexicographic min cost flows in evacuation
    modeling. Nav Res Logist 34:487–503
Jarvis JJ, Ratliff HD (1982) Some equivalent objectives for dynamic network flow problems.
    Manag Sci 28(1):106–108

Chapter 4
A Relaxation-Based Heuristic Approach for the Mixed-Integer Evacuation Model

A basic urban evacuation model is presented in Chap. 3 with the extensions of the possibility of waiting and the reassignment of the traffic routing only to one-way streets. In this chapter a heuristic solution approach will be presented for the urban evacuation model which relaxes some constraints and integrality in a first step. In a following step based on a solution of the relaxation, feasibility is generated for the considered urban evacuation model. We will consider the two-way case (compare Bretschneider and Kimms 2011) and the one-way case. The two-way case is more difficult to handle, so it is considered first. The one-way case can be derived from the two-way case. The approach will be explained for the urban evacuation model without waiting presented in Sect. 3.2 with the objective function (3.1) that minimizes of the average evacuation time. For the variation of the evacuation model with waiting on arcs the relaxation approach can be applied analogously.

4.1 A Relaxation-Based Heuristic Solution Approach

We developed a relaxation-based approach to generate feasible solutions of the urban evacuation model presented in Sect. 3.2 based on a solution of a relaxation of the urban evacuation model. The objective function considered is the average evacuation time for an evacuee which is to be minimized. The relaxation-based approach is divided into three steps:

Stage 1. Solve the relaxation of the basic evacuation model optimally.

Stage 2. Execute the adjustment approach: Adjustment of the ℓ-variables such that they are feasible subject to the constraints of the urban evacuation model (Sect. 3.2).

Stage 3. Solve the urban evacuation model with fixed ℓ-variables achieved by the adjustment approach: Computation of the inflow variables x and the value of the objective function.

S. Bretschneider, *Mathematical Models for Evacuation Planning in Urban Areas*,
Lecture Notes in Economics and Mathematical Systems 659,
DOI 10.1007/978-3-642-28759-6_4, © Springer-Verlag Berlin Heidelberg 2013

The relaxed urban evacuation model is the urban evacuation model without "Coupling constraints" (3.14), "Prohibition of crossing conflicts I, II, III" (3.15)–(3.17) and without the associated variables y needed to prohibit the assignment of a positive number of lanes of certain left-turn or straight forward arcs within intersections. Furthermore the variables $\ell(j,l,i,h) \in \mathbb{N}_0$ describing the number of lanes of arcs $(j,l,i,h) \in \mathcal{A}_{\mathcal{J}} \cup \mathcal{A}_C$ are relaxed to $\ell(j,l,i,h) \in \mathbb{R}_0^+$, $(j,l,i,h) \in \mathcal{A}_{\mathcal{J}} \cup \mathcal{A}_C$. Thus the relaxation contains the following restrictions and decision variables:

- Objective: Minimization of the average evacuation time for every evacuee (3.1)
- Network flow constraints:

    – Standard network flow constraints (3.5)–(3.7)
    – Avoidance of irregular behavior I and II (3.8), (3.9)

- Traffic routing: Lane consistency I and II (3.10)–(3.12), (3.13)
- Capacity Constraints (3.18), (3.19)
- Superfluous Arcs (3.20)–(3.22)
- Decision variables:

    – The number of lanes $\ell(j,l,i,h) \in \mathbb{R}_0^+$ for arcs $(j,l,i,h) \in \mathcal{A}_{\mathcal{J}} \cup \mathcal{A}_C$ representing the street network.
    – The inflow $x(j,l,i,h,t) \in \mathbb{R}_0^+$ for arcs in $(j,l,i,h) \in \mathcal{A}$.

The solution of the relaxed evacuation model does not need to be feasible for the urban evacuation model therefore the solution has to be adjusted properly in Stage 2. The infeasibilities that might occur are the non-integrality of some ℓ's, or the crossing of some lanes within intersections. Pseudocode 1 gives a sketch of the adjustment approach of Stage 2. All steps of the pseudocode will be explained in

---

Pseudocode 1 Overview of the adjustment approach of Stage 2

Step 0:
If there is no significant cumulative flow on arc $(i,h,j,l) \in \mathcal{A}_C \cup \mathcal{A}_{\mathcal{J}}$ then set $\ell(i,h,j,l) := 0$

repeat
    for $i \in \mathcal{J}: \alpha_i \geq 2$ **do**
        Step 1: Generate integrality of $\ell(i,h,i,j)$ for all $(i,h,i,j) \in \mathcal{A}_C: (i,j) \in \mathcal{D}$
        Step 2: Generate integrality of $\ell(i,h,i,j)$ for all $(i,h,i,j) \in \mathcal{A}_C: (i,j) \notin \mathcal{D}$
        repeat
            Step 3.1: Correct crossing conflicts of two straight forward arcs that are prohibited by constraint (3.15)
            Step 3.2: Correct crossing conflicts of directions having a node in common and are prohibited by constraints (3.16) and (3.17)
        until no crossing conflict within intersection i occurs
        Step 3.3: Generate integrality of $\ell(i,h,i,l)$ for all $(i,h,i,l) \in \mathcal{A}_{\mathcal{J}}$
    end for
until the ℓ-variables for every intersection are feasible

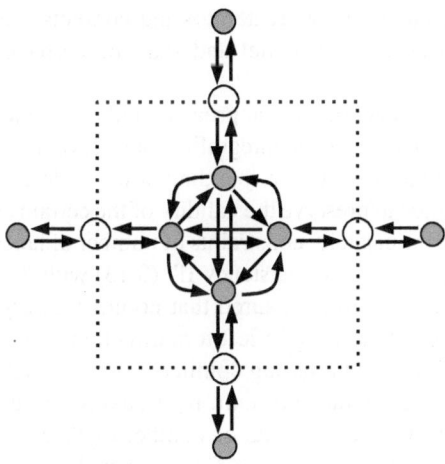

Fig. 4.1 Investigated arcs of an intersection in Steps 1 through 3

Sects. 4.1.1 and 4.1.2. In Sect. 4.1.1 preliminary considerations of the steps will be discussed and in Sect. 4.1.2 the adjustment algorithm will be explained in detail.

In Step 1 through Step 3.3 of the adjustment procedure, every intersection and adjoining arcs are considered separately, that means the ℓ-variables of arcs within the intersection and arcs that adjoin to an intersection node of the investigated intersection are considered. For example, the ℓ-variables of the arcs within the dotted area in Fig. 4.1 are taken into account if the intersection at center is investigated. Steps 1 through 3.3 are repeated for every intersection $i \in \mathcal{J}$ (with $\alpha_i \geq 2$) until feasibility regarding the constraints of the urban evacuation model of the ℓ-variables of all intersections can be guaranteed. During Steps 2 and 3.2 it is possible that new crossing conflicts are created when ensuring that it is possible to reach a destination from every source. If this case occurs, Steps 1 through 3.3 have to be applied to the corresponding intersection again (the outer repeat-until loop). If there are intersections with $\alpha = 2$, where no crossing conflict exists, or $\alpha = 3$, where no crossing conflict of two straight forward arcs exists, then Steps 1, 2 and 3.3 and Steps 1, 2, 3.2 and 3.3 are responsible to generate feasibility for the corresponding intersection, respectively. To describe the adjustment approach the following definitions are used: Node $(i, h) \in \mathcal{N}_{\mathcal{J}}$ of intersection i is called an *exit* of intersection i if $\sum_{h_1 \in \{1, \dots \alpha_i\} \setminus \{h\}} \lambda(i, h_1, i, h) > 0$ and node $(i, h) \in \mathcal{N}_{\mathcal{J}}$ is called an *entrance* of intersection i if $\sum_{h_1 \in \{1, \dots \alpha_i\} \setminus \{h\}} \lambda(i, h, i, h_1) > 0$. A node of an intersection can be an entrance and an exit simultaneously. Remember that node $(i, j) \in \mathcal{S} \cup \mathcal{D}$ and $(j, i) \in \mathcal{S} \cup \mathcal{D}$ stand for the same node.

4.1.1   Preliminary Considerations

The basis for the heuristic adjustment approach is the solution of the relaxation of the urban evacuation model. The values of the decision variables of the number of lanes may be fractional and may create crossing conflicts. These are infeasibility points for the urban evacuation model and shall be corrected with the heuristic adjustment approach.

In Steps 2 and 3.2 of the heuristic approach number of lanes are rounded down, rounded up or set to zero to achieve integrality (Step 2) or are set to zero to delete a crossing conflict (Step 3.2). After the rounding down or deleting, additional operations have to be executed to preserve the validity of the certain constraints. In Step 2, we want to guarantee the capacity constraints of number of lanes "Lane Consistency I" (3.11), the constraint "Lane Consistency II" (3.13) with $\delta = 1$ that restricts the merging of lanes and it has to be assured that no unnecessary blind alley is built. In Step 3.2 the crossing conflicts of a left turn direction with another left turn or a straight forward direction and crossing conflicts of two straight forward directions having the same node as entrance or exit, respectively, are deleted while guaranteeing as well the capacity constraints of number of lanes "Lane Consistency I" (3.10) and (3.11), the constraint "Lane Consistency II" (3.13) and no dead-end building.

In Step 3.1 crossing conflicts of two straight forward arcs where the arcs have no node in common are eliminated with reallocating number of lanes. The constraints "Lane Consistency I" (3.11) and "Lane Consistency II" (3.13) have to stay valid after these operations. The constraint "Lane Consistency II" (3.13) with $\delta = 1$ restricts the merging of lanes of the very same direction and can be stated as follows:

$$\ell(i,h,i,l) \leq \ell(i,l,i,j) \tag{4.1}$$

$$\text{for all } (i,h,i,l) \in \mathcal{A}_\mathcal{J}; (i,l,i,j) \in \mathcal{A}_C$$

Before describing the general case we are going to give simple examples and preliminary considerations. In Step 2 the integrality of the number of lanes of arcs in $\mathcal{A}_C$, i.e. arcs that connect intersections and sources or destinations, is built. Considerations 1–3 are concerned with Step 2.

Consideration 1: Rounding Down the Number of Lanes of an Arc in $\mathcal{A}_C$
Emerging from an Exit of an Intersection I

For the preliminary consideration we consider intersection i that can be viewed in Fig. 4.2a. The arcs depict the directions that have a positive number of lanes in the solution of the relaxation. Assume that in this example, the number of lanes of the "connecting" arc $(i,3,i,j)$ that emerges from exit $(i,3)$ of intersection i has to be rounded down to achieve integrality and to preserve the capacity constraint (3.11) regarding the number of lanes of street segment described by arcs $(i,3,i,j)$ and $(i,j,i,3)$. Let the rounded down value be positive, i.e. $\ell(i,3,i,j) > 0$. The rounding down is depicted in Fig. 4.2a by the thin line of the corresponding arc $(i,3,i,j)$.

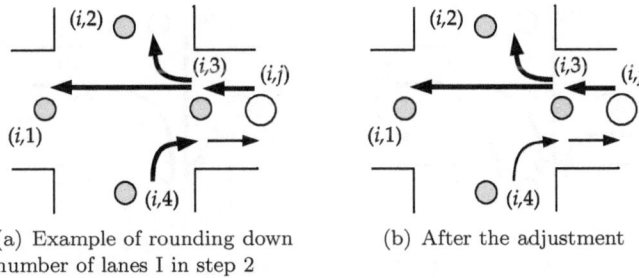

(a) Example of rounding down   (b) After the adjustment
number of lanes I in step 2

Fig. 4.2 Example to consideration 1 concerning the rounding down of number of lanes in Step 2

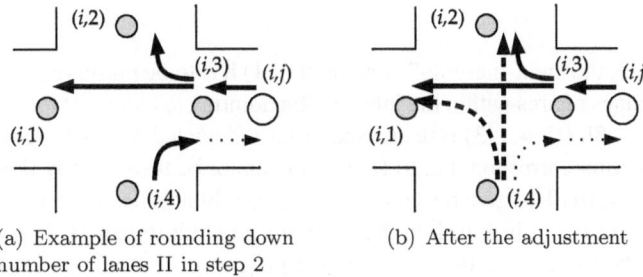

(a) Example of rounding down   (b) After the adjustment
number of lanes II in step 2

Fig. 4.3 Example to consideration 2 concerning the rounding down of number of lanes I in Step 2

In this case, we have to assure that no merging of lanes of a turn within the intersection occurs such that constraint (4.1) stays valid (i.e. $\ell(i, h, i, 3) \leq \ell(i, 3, i, j)$, $h \in \{1, 2, 4\}$). In the example in Fig. 4.2a the number of lanes $\ell(i, 4, i, 3)$ of turn $(i, 4, i, 3)$ has to be rounded down, too, to avoid that $\ell(i, 4, i, 3)$ is greater than $\ell(i, 3, i, j)$ what would be a merging. In Fig. 4.2b the line width of the right turn arc $(i, 4, i, 3)$ is decreased to the one of the arc $(i, 3, i, j)$. The decreasing of the line width depicts the rounding down.

A conclusion of this consideration is that after rounding down the number of lanes of an arc $(i, l, i, j) \in \mathcal{A}_C$ with $(i, j) \in \mathcal{S} \cup \mathcal{D}$ (i.e. in the same time (i, l) is an exit of intersection i, $(i, l) \in \mathcal{N}_\mathcal{J}$) the number of lanes of arcs $(i, h, i, l) \in \mathcal{A}_\mathcal{J}$ of intersection i has to be controlled and probably corrected.

Consideration 2: Rounding Down the Number of Lanes of an Arc in $\mathcal{A}_C$
Emerging from an Exit of an Intersection II

Figure 4.3 illustrates the next example. Again the number of lanes of the "connecting" arc $(i, 3, i, j)$ that emerges from the exit $(i, 3)$ of intersection i have to be rounded down. Let the rounded down value in this case be zero, i.e. $\ell(i, 3, i, j) = 0$. The thin dotted line of arc $(i, 3, i, j)$ shown in Fig. 4.3a depicts that the value of number of lanes is zero.

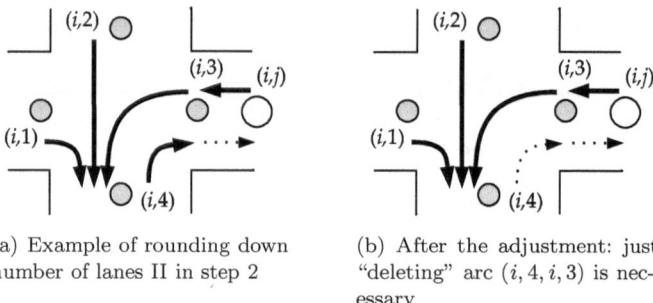

(a) Example of rounding down
number of lanes II in step 2

(b) After the adjustment: just
"deleting" arc $(i, 4, i, 3)$ is necessary

Fig. 4.4 Example to consideration 2 concerning the rounding down of number of lanes II in Step 2

First of all, the "non-merging" constraint (4.1) has to be preserved again, i.e. the number of lanes of arcs within the intersection pointing to exit $(i, 3)$ have to be set to zero. In Fig. 4.3b $\ell(i, 4, i, 3)$ is decreased to zero (depicted with a dotted line). Then the street section corresponding to $(i, 3, i, j)$ cannot be traversed by flow (because $\ell(i, 3, i, j) = 0$). The generation of unnecessary blind alleys has to be avoided: In the given example in Fig. 4.3a the only arc pointing to exit $(i, 3)$ is $(i, 4, i, 3)$. That is, in the same time, the only arc emerging from entrance $(i, 4)$. The source next to entrance $(i, 4)$ might possibly be cut off from a way to a destination because the arcs $(i, 4, i, 3)$ and $(i, 3, i, j)$ can no longer used by flow. The "prior" value of $\ell(i, 4, i, 3)$ is reallocated as much as possible to other exit nodes like $(i, 1)$ and/or $(i, 2)$ of the intersection to preserve from building a blind alley and to "conserve" allocated capacity within the intersection. I.e. the value $\ell(i, 4, i, 3)$ is reallocated in any case (also if $\ell(i, 4, i, 1) > 0$ or $\ell(i, 4, i, 2) > 0$) if possible. This means that fractions of the value $\ell(i, 4, i, 3)$ of the number of lanes can be reassigned to the number of lanes of the arcs $(i, 4, i, 1)$ and/or $(i, 4, i, 2)$ highlighted with dashed lines in Fig. 4.3b. The numbers of lanes after the reallocation have to assure the capacity constraints of number of lanes "Lane Consistency I" (3.10) and constraints "Lane Consistency II" (3.13) for all corresponding δ.

Sometimes it is not possible to find another arc with another exit for the reallocation of the number of lanes. Such a case is illustrated in Fig. 4.4a. The number of lanes of the arc $(i, 4, i, 3)$ should be added to another arc emerging from entrance $(i, 4)$ and pointing to an exit $\neq (i, 3)$ of intersection i. But all arcs with a positive number of lanes point to the node $(i, 4)$, the tail of arc $(i, 4, i, 3)$. Hence there is no other exit available than $(i, 4)$. In this case, the number $\ell(i, 4, i, 3)$ of lanes of $(i, 4, i, 3)$ can just be set to zero (see Fig. 4.4b) without further operations. The flow traveling the arcs $(i, 1, i, 4)$, $(i, 2, i, 4)$ and $(i, 3, i, 4)$ have to reach a destination using another way and so will the amount of flow that traveled arc $(i, 4, i, 3)$ in the solution of the relaxation.

It is possible that after the reallocation of number of lanes new crossing conflicts occur. These are corrected in subsequent Steps 3.1 and 3.2.

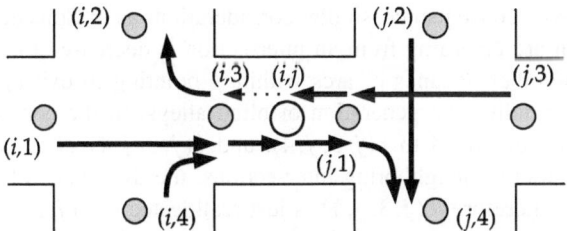

(a) Example of decreasing the number of lanes of an arc
in $\mathcal{A}_C$ pointing to an entrance of an intersection to zero

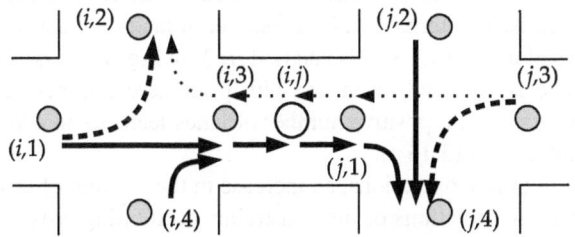

(b) Intersections i and j after the adjustment

Fig. 4.5 Example of consideration 3 concerning the decreasing the number of lanes of an arc in $\mathcal{A}_C$ pointing to an entrance of an intersection to zero

Consideration 3: Decreasing the Number of Lanes of an Arc in $\mathcal{A}_C$ Pointing to an Entrance of an Intersection to Zero

Consider the example illustrated in Fig. 4.5. Figure 4.5a gives the initial scenario: arc $(i, j, i, 3)$ in $\mathcal{A}_C$ pointing to the entrance of an intersection is set to zero (the arc with the dotted line). Then intersection i investigated in the actual iteration and the appropriate neighboring intersection j have to be controlled.

Intersection i: The value of the number of lanes of all arcs within intersection i emerging from $(i, 3)$ can be set to zero because no flow can emerge from $(i, 3)$. If there is an arc, like $(i, 3, i, 2)$, emerging from $(i, 3)$ that is the only arc with a positive number of lanes pointing to an exit, here $(i, 2)$, then we can set the value of number of lanes of another arc pointing to the exit $(i, 2)$ to the "prior" $\ell(i, 3, i, 2)$ to "offer" the exit to another direction. The available number of lanes (constraints (3.10)) has to be regarded. See for an illustration Fig. 4.5b: $\ell(i, 3, i, 2)$ is decreased to zero (depicted with a thin dotted arc) and the value $\ell(i, 1, i, 2)$ of number of lanes of arc $(i, 1, i, 2)$ has been set to the prior value $\ell(i, 3, i, 2)$ of $(i, 3, i, 2)$ (the increase of number of lanes is depicted as a dashed bold arc).

Neighboring intersection j: If $\ell(i, j, i, 3)$ is rounded down to zero then the neighboring intersection, intersection j in Fig. 4.5a on the right, may be effected. If $\ell(i, j, i, 3) = 0$, then no flow can travel arc $(i, j, i, 3)$ and therefore no flow can travel arc $(j, 1, i, j)$ as well, because no turnaround is allowed. If entering arc $(j, 1, i, j)$, the flow could not leave node (i, j). Hence $\ell(j, 1, i, j)$ can be set to zero, as well. As a consequence, similar considerations as in Consideration 2 occur: the value of an arc emerging from an intersection is decreased to zero. Therefore all values of number of lanes of arcs within j pointing to exit $(j, 1)$ can be set to zero while avoiding the generation of blind alleys. In the considered example $\ell(j, 3, j, 1)$ may be added to $\ell(j, 3, j, 4)$, and $\ell(j, 3, j, 1)$ is be set to zero, see Fig. 4.5b. Considering neighboring intersections, the avoidance of blind alleys is the main goal. Therefore, $\ell(j, 3, j, 1)$ is just reallocated to $\ell(j, 3, j, 4)$, if no other arc within intersection j emerging from entrance $(j, 3)$ has a positive ℓ-value (see the given example) and at most the value one is added to $\ell(j, 3, j, 4)$.

The feasibility of capacity constraints (3.10) concerning the number of lanes within intersections have to be taken into account. "Lane Consistency II" (3.13) is preserved, because in intersection i the number of lanes are just "exchanged" (the number of lanes of the arcs $(i, 3, i, 2)$ and $(i, 1, i, 2)$ in Fig. 4.5), and in intersection j at most one lane is assigned to an arc with prior zero number of lanes (i.e. an additional direction with a positive number of lanes leads to an additional allowed lane (see formulation of (3.13))).

After executing these operations, an increase in the number of crossing conflicts may occur but these violations of the constraints prohibiting crossing conflicts are ignored in Step 2 and are corrected in Step 3 of the heuristic algorithm.

Consideration 4: Deleting Crossing Conflicts Corresponding to Constraint (3.15) in Step 3.1

Crossing conflicts of two straight forward arcs that have no node in common are repaired in Step 3.1. Let (i, h_1, i, h_2) and (i, l_1, i, l_2) be two straight forward arcs that cross each other and build a crossing conflict, i.e. $\ell(i, h_1, i, h_2) > 0$ and $\ell(i, l_1, i, l_2) > 0$. Figure 4.6 illustrates such a crossing conflict. The crossing conflict (i, h_1, i, h_2) and (i, l_1, i, l_2) is corrected if $\ell(i, h_1, i, h_2) = 0$ or $\ell(i, l_1, i, l_2) = 0$.

As many lanes as necessary are reallocated from (i, h_1, i, h_2) to (i, h_1, i, l_2) and from (i, l_1, i, l_2) to (i, l_1, i, h_2) such that the crossing conflict is deleted (if possible), the constraint "Lane Consistency I" (3.10) for intersection arcs and the constraint "Lane Consistency II" with $\delta = 1$ (see (4.1)) is not violated. That means, it is possible to add the minimum of the values $\ell(i, h_1, i, h_2)$, $\ell(i, l_1, i, l_2)$, $\mathcal{L}(i, h_1, i, l_2) - \ell(i, h_1, i, l_2) - \ell(i, l_2, i, h_1)$ (preserving the feasibility of (3.10)), $\ell(i, l_2, i, j_1) - \ell(i, h_1, i, l_2)$ (preserving the feasibility of (4.1)) with $(i, l_2, i, j_1) \in \mathcal{A}_C$ to the number of lanes $\ell(i, h_1, i, l_2)$ of arc (i, h_1, i, l_2). The reallocation of number of lanes is analog for $\ell(i, l_1, i, h_2)$. See again Fig. 4.6 for a generic scheme of correcting a crossing conflict. The arcs show directions with a positive number of lanes.

After the reallocation of number of lanes, the constraint "Lane Consistency II" (3.13) has to remain valid for the exits (i, l_2) and (i, h_2) where $\delta = 1, \ldots, \alpha_i - 1$. If the constraint "Lane Consistency II" (3.13) for a certain δ does not stay valid after the manipulations of $\ell(i, h_1, i, h_2)$, $\ell(i, l_1, i, l_2)$, $\ell(i, h_1, i, l_2)$ and $\ell(i, l_1, i, h_2)$, it has to be corrected by subtracting the appropriate number of lanes from $\ell(i, h_1, i, l_2)$ and $\ell(i, l_1, i, h_2)$, respectively. See Fig. 4.7 for an example

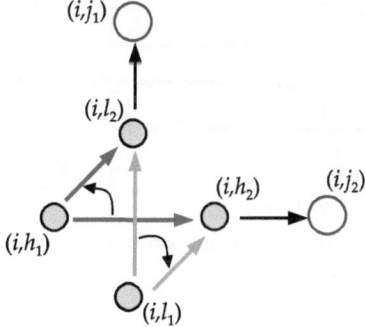

Fig. 4.6 Illustration of consideration 4: generic scheme of repairing a crossing conflict of two straight forward arcs

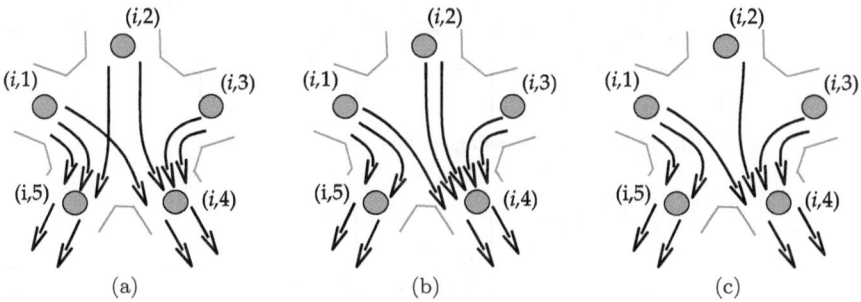

Fig. 4.7 (**a**) Intersection with one crossing conflict (**b**) Intersection after deleting the crossing conflict (**c**) Rebuilding validity of lane consistency II (3.13)

of rebuilding the validity of "Lane Consistency II" (3.13). Here every arc depicts one lane. Figure 4.7a shows the solution of the ℓ-variables of the relaxation within intersection i. Directions $(i, 1, i, 4)$ and $(i, 2, i, 5)$ build a crossing conflict. In Fig. 4.7b the crossing conflict is repaired with the above described procedure, i.e. the values of number of lanes $\ell(i, 1, i, 4)$, $\ell(i, 1, i, 5)$, $\ell(i, 2, i, 5)$ and $\ell(i, 2, i, 4)$ are recalculated. But "Lane Consistency II" (3.13) is violated in node $(i, 4)$ for example for $\delta = 3$: $\sum_{(i,l,i,4)\in A_{\mathscr{J}}} \ell(i, l, i, 4) > \ell(i, 4, i, j_1) + \delta - 1$ (i.e. $5 > 2 + 3 - 1$). In Fig. 4.7c the validity of constraint (3.13) is rebuilt by recalculating $\ell(i, 2, i, 4)$ with $\ell(i, 2, i, 4) := \ell(i, 2, i, 4) - (\sum_{(i,l,i,4)\in A_{\mathscr{J}}} \ell(i, l, i, 4) - (\ell(i, 4, i, j_1) + \delta - 1)) = 2 - (5 - 4) = 1$.

If after these manipulations both values $\ell(i, h_1, i, h_2)$ and $\ell(i, l_1, i, l_2)$ are still greater than zero, the crossing conflict is not removed and the smaller one is set to zero to correct the crossing conflict.

Consideration 5: Deleting Crossing Conflicts Corresponding to Constraints (3.16) and (3.17)

In Step 3.2 the crossing conflicts of left or straight forward arcs having the same node as entrance or exit, respectively, are corrected while the capacity constraint of

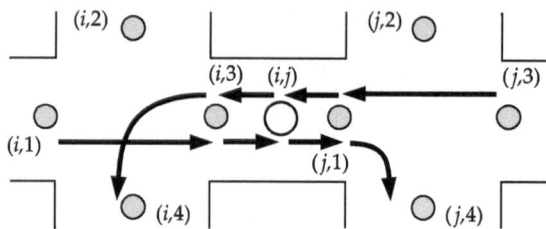

(a) Example of a crossing conflict of a left turn and straight forward arc

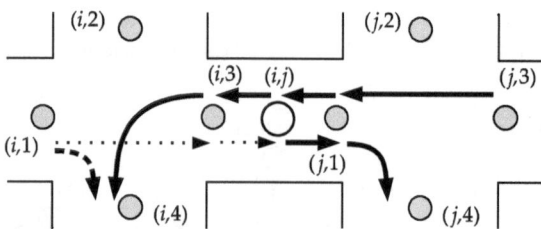

(b) Deleting the crossing conflict with $\ell(i, 1, i, 3) := 0$

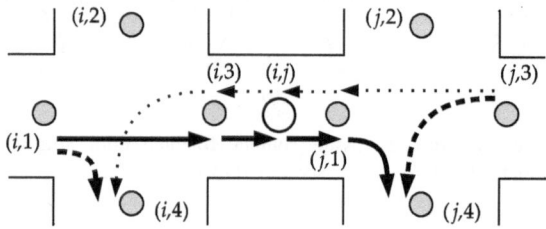

(c) Deleting the crossing conflict with $\ell(i, 3, i, 4) := 0$

Fig. 4.8 Illustration of consideration 5 concerning deleting crossing conflicts

number of lanes "Lane Consistency I" (3.10), the constraint "Lane Consistency II" (3.13) and the avoidance of building a blind alley have to be guaranteed. See Fig. 4.8 for an illustration.

Crossing conflicts of arcs where the head of an arc is equal to the tail from the other arc are corrected in the following way: a value of number of lanes of exactly one arc is set to zero. With that operation the crossing conflict disappears and we try to add the deleted value to a certain arc without violating constraints (3.10) and (3.13). Decreasing the value of number of lanes to zero may lead to complications with respect to feasibility and further operations has to be executed.

In Fig. 4.8a, an example of a crossing conflict of left turn arc $(i, 3, i, 4)$ and straight forward arc $(i, 1, i, 3)$ is given. Either $\ell(i, 1, i, 3)$ or $\ell(i, 3, i, 4)$ is set to zero. In both cases at most $\ell(i, 1, i, 3)$ or $\ell(i, 3, i, 4)$ is added to $\ell(i, 1, i, 4)$ such that constraints (3.10) and (3.13) remain feasible, respectively.

If the value of number of lanes of the straight forward arc $(i, 1, i, 3)$ is set to zero and if no other lanes leave intersection through exit $(i, 3)$, then $\ell(i, 3, i, j)$ can be set to zero. See Fig. 4.8b: Just to arc $(i, 1, i, 3)$ was a positive number of lanes assigned, so after setting $\ell(i, 1, i, 3) := 0$ no flow can enter arc $(i, 3, i, j)$ and $\ell(i, 3, i, j) > 0$ is no longer needed.

If the number of lanes of the left turn arc $(i, 3, i, 4)$ is set to zero (see Fig. 4.8c) and the left turn arc is the only arc with a positive number of lanes, then the value of number of lanes of arc $(i, j, i, 3)$ can be set to zero, because no flow can leave node $(i, 3)$ due to the prohibition of turnarounds. That leads us to the case considered in Consideration 2: the neighboring intersection j has to be controlled and the appearance of certain blind alleys has to be avoided. Figure 4.8c shows the example of Fig. 4.8a after further operations that have to be executed to preserve feasibility. Bold dashed arcs depict the arcs with increased number of lanes, thin dotted arcs depict directions with the number of lanes set to zero and the bold arcs depict arcs that have not been changed.

4.1.2 Adjustment Algorithm

Input of the adjustment approach is an optimal solution of the relaxed urban evacuation model. The values of number of lanes, the ℓ-values, of the arcs corresponding to the street network have to be adjusted such they are feasible for the urban evacuation model. An overview of the adjustment algorithm is given in Pseudocode 1 (on page 68). In the following the steps of the algorithm are described in more detail.

Step 0:

The assignment of a positive number of lanes to an arc that is traversed by only a small total amount of flow may waste "lane capacity". Therefore every ℓ-value of an arc that is not traversed by a total amount of flow of at least χ is set to zero, i.e. $\ell(i, h, j, l) := 0$ for all $(i, h, j, l) \in \mathcal{A}_C \cup \mathcal{A}_{\mathcal{J}}$ with $\sum_{t \in T} x(i, h, j, l, t) < \chi$.

For every intersection $i \in \mathcal{J}$ with $\alpha_i \geq 2$ the Step 1 until Step 3.2 are executed:

Step 1:

All values $\ell(i, h, i, j)$ with $(i, h, i, j) \in \mathcal{A}_C$ with $(i, j) \in \mathcal{D}$ are rounded up, i.e. $\ell(i, h, i, j) := \lceil \ell(i, h, i, j) \rceil$.

Step 2:

In Step 2 the number of lanes of arcs that connect intersection i with a source are considered. Round the values $\ell(i, h, i, j)$ and $\ell(i, j, i, h)$, where $(i, h, i, j) \in \mathcal{A}_C$

Pseudocode 2 Rounding decisions in Step 2

---

for $(i, h, i, j) \in \mathcal{A}_C : (i, j) \in \mathcal{S}$ **do**

    if $\lceil \ell(i, h, i, j) \rceil + \lceil \ell(i, j, i, h) \rceil \leq \mathcal{L}(i, h, i, j)$ **then**

      $\ell(i, h, i, j) := \lceil \ell(i, h, i, j) \rceil$

      $\ell(i, j, i, h) := \lceil \ell(i, j, i, h) \rceil$

    else

      if decimal place value of $\ell(i, h, i, j)$ < decimal place value of $\ell(i, j, i, h)$ **then**

        $\ell(i, h, i, j) := \lfloor \ell(i, h, i, j) \rfloor$

        $\ell(i, j, i, h) := \lceil \ell(i, j, i, h) \rceil$

      else

        $\ell(i, h, i, j) := \lceil \ell(i, h, i, j) \rceil$

        $\ell(i, j, i, h) := \lfloor \ell(i, j, i, h) \rfloor$

      end if

    end if

end for

---

with $(i, j) \in \mathcal{S}$, up if possible (i.e. without violating the capacity constraint of number of lanes (3.11)), otherwise round the ℓ-value with the lesser decimal place value down and the other one up, see Pseudocode 2. After these operations there are three cases:

Case 1: $\ell(i, h, i, j) := \lceil \ell(i, h, i, j) \rceil$ and $\ell(i, j, i, h) := \lceil \ell(i, j, i, h) \rceil$

Then, nothing needs to be done. The capacity constraint "Lane consistency I" (3.11) of number of lanes is preserved (there is no modification of the number of lanes of arcs within intersections) and constraint "Lane consistency II" (3.13) cannot not be violated ($\ell(i, h, i, j) + \delta - 1 \leq \lceil \ell(i, h, i, j) \rceil + \delta - 1$, where $\ell(i, h, i, j)$, $(i, h, i, j) \in \mathcal{A}_C$, is the value before the manipulation).

Case 2: $\ell(i, h, i, j) := \lfloor \ell(i, h, i, j) \rfloor$ and $\ell(i, j, i, h) := \lceil \ell(i, j, i, h) \rceil$

Feasibility subject to constraint "Lane consistency II" (3.13) restricting the number of mergings within an intersection, possibly, has to be rebuilt and the generation of unnecessary blind alleys has to be avoided. The rounded down value $\ell(i, h, i, j)$ may lead to infeasibilities. The following two cases arise:

$\ell(i, h, i, j) > 0$ The merging of lanes of the same direction leading to exit (i, h) is no longer guaranteed. That property describes constraint "Lane Consistency II" (3.13) with $\delta = 1$, see constraint (4.1) for a reformulation. The number of lanes of arcs within intersection i that point to exit (i, h) have to be controlled and if necessary changed. If constraint "Lane Consistency II" with $\delta = 1$ (4.1) is violated for an arc $(i, a, i, h) \in \mathcal{A}_{\mathcal{J}}$ feasibility has to be readjusted properly by following Consideration 1: If $\ell(i, a, i, h) > \ell(i, h, i, j)$ then let $\ell(i, a, i, h) := \ell(i, h, i, j)$ for all $(i, a, i, h) \in \mathcal{A}_{\mathcal{J}}$.

$\ell(i, h, i, j) = 0$ If $\ell(i, h, i, j)$ is zero after rounding down, no flow can exit intersection i through the exit (i, h). That means in the same time that no flow can leave intersection i using arcs $(i, a, i, h) \in \mathcal{A}_{\mathcal{J}}$ even if $\ell(i, a, i, h) > 0$. See Fig. 4.9 for an illustration. The dotted arc depicts that $\ell(i, h, i, j)$ has been decreased to zero in Step 2 with the procedure described in Pseudocode 2. All other arcs have a positive number of lanes.

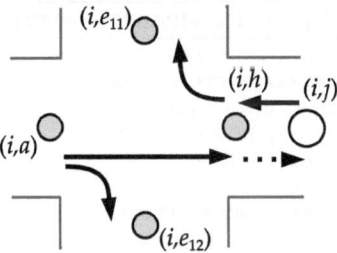

Fig. 4.9 Illustration of Step 2 Case 2 with $\ell(i, h, i, j) = 0$

Another exit than (i, h) of i has to be used to traverse intersection i. Consideration 2 helps here. The number of lanes of intersection arcs $(i, a, i, h) \in \mathcal{A}_{\mathcal{J}}$ with $\ell(i, a, i, h) > 0$ is reallocated if possible to other arcs in the intersection. The amount $\ell(i, a, i, h) > 0$ is reallocated completely, if possible, to $\ell(i, a, i, e_1)$ where $(i, a, i, e_1) \in \mathcal{A}_{\mathcal{J}}$ and $e_1 \neq h$ is an exit. In the example, given in Fig. 4.9, (i, e_{11}) and (i, e_{12}) are exits of intersection i. Node (i, e_1) is an exit if $\sum_{(i,l,i,e_1) \in \mathcal{A}_{\mathcal{J}}} \ell(i, l, i, e_1) > 0$. The reallocation has to take the constraint "Lane Consistency I"(3.10) for arcs within intersection i and "Lane Consistency II" (3.13) into account. The reallocation of lanes is needed to assure that from every source a way leading out of the evacuation area to a sink exists.

For all exits (i, e_1) (with $e_1 \neq h$) of intersection i the procedure described in Pseudocode 3 is executed until the total amount of $\ell(i, a, i, h)$ is reallocated or all exits are investigated. The order of the investigation of the exits is not optimized, they are considered in the order of the number of exits.

Value $\Delta_1 := \min\{\ell(i, a, i, h), \mathcal{L}(i, a, i, e_1) - \ell(i, a, i, e_1) - \ell(i, e_1, i, a),$ $\ell(i, e_1, i, j_1) - \ell(i, a, i, e_1)\}$ (Line 3) gives the maximal number of lanes that can be added to direction (i, a, i, e_1) without violating "Lane Consistency I" (3.10) and without violating "Lane Consistency II" (3.13) for $\delta = 1$. The value of number of lanes that leads to the infeasibility of "Lane Consistency II" (3.13) is determined by $\Delta_2 = (\sum_{(i,a_1,i,e_1) \in A^*} \ell(i, a_1, i, e_1)) - (\ell(i, e_1, i, j_1) + |A^*| - 1)$ (Line 8), where A^* is a subset of intersection arcs that point to the considered exit (i, e_1) and includes the arc (i, a, i, e_1). In Line 4 of Pseudocode 3 the maximal value of number of lanes is added to the new direction (i, a, i, e_1) and in Line 5 the remaining number of lanes that can be reallocated to the next arc emerging from (i, a) is determined with updating $\ell(i, a, i, h)$. The for-loop beginning in Line 1 is repeated until there are no longer lanes to be reallocated (Line 13–15) or all possible exits are considered. $\ell(i, a, i, h)$ is set to zero in any case (Line 17).

The case that there is no exit of intersection i the number of lanes $\ell(i, a, i, h)$ can be reallocated, discussed in Consideration 2, is captured with the above described algorithm as well: $\ell(i, a, i, h)$ is set to zero.

Pseudocode 3 Case 2 in Step 2 of the adjustment heuristic

1: **for** $(i, a, i, h) \in \mathcal{A}_{\mathcal{J}}$ with $\ell(i, a, i, h) > 0$ **do**
2:    **for** $(i, e_1) \neq (i, h)$ is an exit of intersection i **do**
3:      $\Delta_1 := \min\{\ell(i, a, i, h), \mathcal{L}(i, a, i, e_1) - \ell(i, a, i, e_1) - \ell(i, e_1, i, a), \ell(i, e_1, i, j_1) - \ell(i, a, i, e_1)\}$ with $(i, e_1, i, j_1) \in \mathcal{A}_C$
4:      $\ell(i, a, i, e_1) := \ell(i, a, i, e_1) + \Delta_1$
5:      $\ell(i, a, i, h) := \ell(i, a, i, h) - \Delta_1$
6:      **for** $\delta = 1, \ldots, \alpha_i - 1$ and all subsets A^* of $\{(i, k_1, i, k_2) \in \mathcal{A}_{\mathcal{J}} : k_2 = e_1\}$ that include arc (i, a, i, e_1) **do**
7:        **if** $\sum_{(i, a_1, i, e_1) \in A^*} \ell(i, a_1, i, e_1) > \ell(i, e_1, i, j_1) + |A^*| - 1$ **then**
8:          $\Delta_2 := (\sum_{(i, a_1, i, e_1) \in A^*} \ell(i, a_1, i, e_1)) - (\ell(i, e_1, i, j_1) + |A^*| - 1)$
9:          $\ell(i, a, i, e_1) := \ell(i, a, i, e_1) - \Delta_2$
10:          $\ell(i, a, i, h) := \ell(i, a, i, h) + \Delta_2$
11:        **end if**
12:      **end for**
13:      **if** $\ell(i, a, i, h) = 0$ **then**
14:        leave the for-loop
15:      **end if**
16:    **end for**
17:    $\ell(i, a, i, h) := 0$
18: **end for**

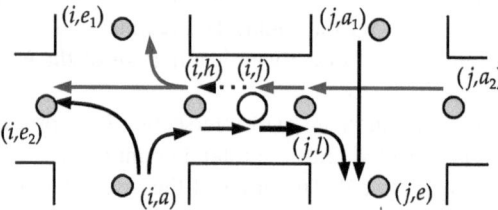

Fig. 4.10 Illustration of Step 2 Case 3 with $\ell(i, j, i, h) = 0$

Case 3: $\ell(i, h, i, j) := \lceil \ell(i, h, i, j) \rceil$ and $\ell(i, j, i, h) := \lfloor \ell(i, j, i, h) \rfloor$
    If $\ell(i, j, i, h)$ is equal to zero after rounding down then the ℓ-values of the intersection investigated in this iteration are considered and as well some ℓ-values of the neighboring intersection j have to be examined. For a first example and primarily considerations, we refer to Consideration 3. For a visualization, we refer to Fig. 4.10. The dotted arc (i, j, i, h) depicts that the number of lanes of (i, j, i, h) has been decreased to zero in Step 2. The solid arcs depict directions with a positive number of lanes. The gray arcs are arcs that will be decreased to zero in further operations.

Controlling intersection i: If there is an arc $(i, h, i, e) \in \mathcal{A}_{\mathcal{J}}$ that is the only arc pointing to an exit (i, e) with a positive number of lanes, the value $\ell(i, h, i, e)$ can be reallocated such that the exit (i, e) is "offered" to a new turn arc. In Fig. 4.10 arc (i, h, i, e_1) demonstrates such an example. The new turn in this example would be (i, a, i, e_1). The value of number of lanes of intersection arcs emerging from entrance (i, h) can be set to zero, i.e: for all $(i, h, i, e) \in \mathcal{A}_{\mathcal{J}}$ set $\ell(i, h, i, e) := 0$. For the example given in Fig. 4.10 that

---

Pseudocode 4 Step 2 Case 3: intersection i

1: **for** $(i, h, i, e) \in \mathcal{A}_{\mathcal{J}}$
   with $\ell(i, h, i, e) > 0$ and $\sum_{(i, a_1, i, e) \in \mathcal{A}_{\mathcal{J}} : a_1 \neq h} \ell(i, a_1, i, e) = 0$ **do**

2:    **for** $(i, a_1) \neq (i, h)$ is an entrance of intersection i **do**

3:        $\ell(i, a_1, i, e) := \min\{\ell(i, h, i, e), \mathcal{L}(i, a_1, i, e) - \ell(i, e, i, a_1)\}$

4:        $\ell(i, h, i, e) := 0$

5:        break

6:    **end for**

7: **end for**

---

means: $\ell(i, h, i, e_1) := 0$ and $\ell(i, h, i, e_2) := 0$. These arcs are pictured as gray arcs in the left intersection.

The procedure for intersection i in Case 3 (see Pseudocode 4) is similar to the procedure in Case 2 (Pseudocode 3) similarly reallocating the number of lanes. In this case, the reallocation is just executed if no other arc than (i, h, i, e) "uses" an exit (i, e) (see Line 1 in Pseudocode 4) because (i, e) will be "unusable" when decreasing $\ell(i, h, i, e)$ to zero in Line 4. To "offer" the exit to another entrance than (i, h), at most $\ell(i, h, i, e)$ is reassigned to an arc (i, a_1, i, e), where (i, a_1) is an entrance of the intersection, while the capacity of number of lanes "Lane Consistency I" (3.10) and "Lane Consistency II" (4.1) have to be preserved. I.e. $\min\{\ell(i, h, i, e), \mathcal{L}(i, a_1, i, e) - \ell(i, e, i, a_1)\}$ number of lanes can be assigned to arc (i, a_1, i, e) (the counter direction has to be considered, see Line 3). Arc (i, a_1, i, e) is the only arc with a positive number of lanes pointing to exit (i, e) ($\ell(i, h, i, e)$ will be decreased to zero afterwards), hence just "Lane Consistency II" (4.1) for $\delta = 1$ needs to be considered. That constraint remains valid for $\ell(i, a_1, i, e)$ because it was for $\ell(i, h, i, e)$.

After the reallocation $\ell(i, h, i, e)$ is set to zero and the for-loop has to be left. Controlling the neighboring intersection j: Consideration 3 underlies the following actions. Let (i, j, j, l) be the appropriate arc in $\mathcal{A}_C$ emerging from source (i, j) between the intersections i and j to intersection j. The values of (i, j, j, l) and (j, l, i, j) can be set to the following values: $\ell(j, l, i, j) := 0$ and $\ell(i, j, j, l) := \mathcal{L}(i, j, j, l)$. For a visualization see Fig. 4.10. The following actions are analog to the ones in Case 2 of Step 2, when the number of lanes that is associated with an arc emerging from an intersection and pointing to a source is decreased to zero, like now $\ell(j, l, i, j)$: It has to be assured that no unnecessary blind alley is built in the neighboring intersection j. For each arc $(j, a, j, l) \in \mathcal{A}_{\mathcal{J}}$ that is the only arc with a positive value of number of lanes emerging from entrance (j, a) (i.e. $\ell(j, a, j, e) = 0$ for all $(j, e) \in \mathcal{A}_{\mathcal{J}}(j)$ with $(j, e) \neq (j, l)$) the number of lanes $\ell(j, a, j, l)$ has to be reallocated to arcs (j, a, j, e) (with $(j, e) \neq (j, l)$ is an exit of j) while taking the constraints "Lane consistency I" (3.10) and "Lane consistency II" (3.13) into account. See for an example Fig. 4.10: the gray arc (j, a_2, j, l) is the only arc with a positive number of lanes pointing to exit (j, l). Node (j, e) is an exit of j if $\sum_{(j, g, j, e) \in \mathcal{A}_{\mathcal{J}}} \ell(j, g, j, e) > 0$.

Considering the neighboring intersection j the following distinction is taken: if either j was not investigated in preceding iterations and feasibility is not guaranteed or if intersection j was already corrected in a preceding iteration and is, in this iteration, feasible for the constraints of the urban evacuation model.

Intersection j has not yet been investigated   In this case simpler operations as taken in Case 2 are executed: the ℓ-value of a randomly found arc (j, a, j, e) (with $(j, e) \neq (j, l)$ is an exit of j) is set to $\min\{\lambda(j, a, j, l), 1, \mathcal{L}(j, a, j, e) - \ell(j, e, j, a)\}$. In Fig. 4.10 the only possibility is to choose arc (j, a_2, j, e) and let $\ell(j, a_2, j, e) := 1$. Adding, at most, one lane to a direction (j, a, j, e) with $\ell(j, a, j, e) = 0$ is always possible without violating constraint "Lane consistency II" (3.13), see the formulation of constraint (3.13) (on page 48).

Intersection j has been investigated   We try to assure that the ℓ-values of intersection j stay feasible subject to the constraints of the urban evacuation model. If there is no other exit than (j, a) then nothing needs to be done. Otherwise, the number of lanes of a randomly chosen arc (j, a, j, e) (with $(j, e) \neq (j, l)$ is an exit of j) that does not lead to a crossing conflict is set to one. If it is not possible to find an arc that does not lead to a crossing conflict then $\min\{\lambda(j, a, j, l), 1, \mathcal{L}(j, a, j, e) - \ell(j, e, j, a)\}$ is assigned to the first found arc (j, a, j, e) (where $(j, e) \neq (j, l)$ is an exit of j) and Steps 1 through Step 3.3 have to be applied to intersection j again.

Step 3

In Step 3 crossing conflicts are repaired and integrality is built within intersection i. Steps 3.1 and 3.2 are repeated until no crossing conflict within the considered intersection i exists. Every crossing conflict is investigated and repaired separately. Step 3.3 assures integrality by observing every exit (i, h) of intersection i separately and assuring the feasibility of "Lane Consistency I" (3.10), the capacity constraint of number of lanes within an intersection and "Lane Consistency II" (3.13).

Step 3.1:

In Step 3.1 crossing conflicts of two straight forward arcs that have no node in common (prohibited by constraint (3.15)) are considered. Let (i, h_1, i, h_2) and (i, l_1, i, l_2) be two straight forward arcs crossing each other and $\ell(i, h_1, i, h_2) > 0$ and $\ell(i, l_1, i, l_2) > 0$. The crossing conflict (i, h_1, i, h_2) and (i, l_1, i, l_2) is corrected if $\ell(i, h_1, i, h_2) = 0$ or $\ell(i, l_1, i, l_2) = 0$. The algorithm to correct a crossing conflict follows Consideration 4: As many lanes as necessary are reallocated from (i, h_1, i, h_2) to (i, h_1, i, l_2) and from (i, l_1, i, l_2) to (i, l_1, i, h_2) while preserving constraints "Lane Consistency I" (3.10) and "Lane Consistency II" (3.13). Figure 4.11 illustrates the deleting of a crossing conflict.

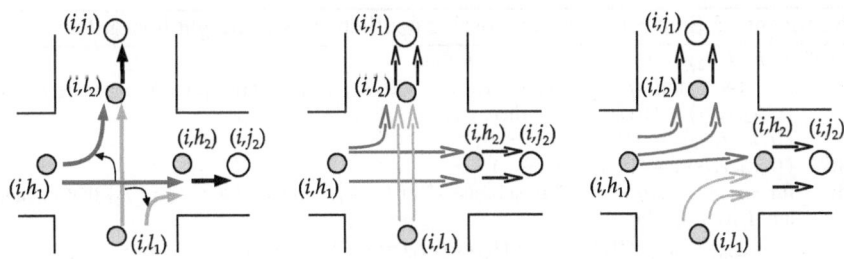

(a) Scheme to delete a crossing conflict of straight forward arcs

(b) Illustration of a crossing conflict of straight forward arcs

(c) Illustration of the number of lanes after deleting a crossing conflict

Fig. 4.11 Illustration of deleting crossing conflicts of two straight forward arcs

For all pairs (i, h_1, i, h_2) and (i, l_1, i, l_2) of straight forward arcs of intersection i that cross the procedure described in Pseudocode 5 has to be executed. A generic scheme to repair a crossing conflict is given in Fig. 4.11a. The arcs depict the considered directions. Lines 2–10 of the algorithm stated in Pseudocode 5 describe the reallocation procedure of $\ell(i, h_1, i, h_2)$ to $\ell(i, h_1, i, l_2)$ (the dark gray arcs in Fig. 4.11a) and Lines 11–19 the procedure to reallocate $\ell(i, l_1, i, l_2)$ to $\ell(i, l_1, i, h_2)$ (the light gray arcs in Fig. 4.11a). The values Δ_1 (Line 2) and Δ_3 (Line 11) are the maximal number of lanes that are possible to add to $\ell(i, h_1, i, l_2)$ and $\ell(i, l_1, i, h_2)$, respectively, without violating constraint "Lane Consistency I" (3.10) and "Lane Consistency II" (3.13) with $\delta = 1$. In Lines 5–10 and Lines 14–19, constraint "Lane Consistency II"(3.13) is controlled and if violated, feasibility is rebuilt for arcs (i, h_1, i, l_2) and (i, l_1, i, h_2), respectively. An example of a crossing conflict is given in Fig. 4.11b and it is repaired in Fig. 4.11c. Every arc depicts a lane in these two figures. The colors orientate on the colors in the generic scheme in Fig. 4.11a.

If, after the reallocation, both arcs leading to the crossing conflict still have a positive number of lanes, the smaller value is set to zero (Lines 20–26 of Pseudocode 5).

Step 3.2:

In Step 3.2 the crossing conflicts of two left turn arcs, a left turn arc and a straight forward arc and crossing conflicts of two straight forward arcs having a tail and head node in common are repaired. These are the crossing conflicts prohibited by constraints "Prohibition of crossing conflicts II, III" (3.16) and (3.17). Let these two arcs be (i, h_1, i, h_2) and (i, l_1, i, l_2) with $h_1 = l_2$ or $l_1 = h_2$. Here we consider the case $h_1 = l_2$, the procedure for case $l_1 = h_2$ is analog. The difference is the particular consideration of arc (i, h_1, i, l_2) for the case $l_1 = h_2$ in the repairing procedure and not arc (i, l_1, i, h_2) as in the presented case $h_1 = l_2$.

Pseudocode 5 Step 3.1 Deleting crossing conflicts of two straight forward arcs

---

1: **if** $\ell(i, h_1, i, h_2) > 0$ and $\ell(i, l_1, i, l_2) > 0$ **then**
2:     $\Delta_1 := \min\{\ell(i, h_1, i, h_2), \ell(i, l_1, i, l_2), \mathcal{L}(i, h_1, i, l_2) - \ell(i, h_1, i, l_2) - \ell(i, l_2, i, h_1),$
        $\ell(i, l_2, i, j_1) - \ell(i, h_1, i, l_2)\}$ where $(i, l_2, i, j_1) \in \mathcal{A}_C$
3:     $\ell(i, h_1, i, h_2) := \ell(i, h_1, i, h_2) - \Delta_1$
4:     $\ell(i, h_1, i, l_2) := \ell(i, h_1, i, l_2) + \Delta_1$
5:     **for** $\delta = 1, \ldots, \alpha_i - 1$ and all subsets A^* of $\{(i, k_1, i, k_2) \in \mathcal{A}_{\mathcal{J}} : k_2 = l_2\}$ that include
        arc (i, h_1, i, l_2) **do**
6:         **if** $\sum_{(i, l, i, l_2) \in A^*} \ell(i, l, i, l_2) > \ell(i, l_2, i, j_1) + |A^*| - 1$ **then**
7:             $\Delta_2 := (\sum_{(i, l, i, l_2) \in A^*} \ell(i, l, i, l_2)) - (\ell(i, l_2, i, j_1) + |A^*| - 1)$
8:             $\ell(i, h_1, i, l_2) := \ell(i, h_1, i, l_2) - \Delta_2$
9:         **end if**
10:     **end for**
11:     $\Delta_3 := \min\{\ell(i, h_1, i, h_2), \ell(i, l_1, i, l_2), \mathcal{L}(i, l_1, i, h_2) - \ell(i, l_1, i, h_2) - \ell(i, h_2, i, l_1),$
        $\ell(i, h_2, i, h_1) - \ell(i, l_1, i, h_2)\}$ where $(i, h_2, i, j_2) \in \mathcal{A}_C$
12:     $\ell(i, l_1, i, l_2) := \ell(i, l_1, i, l_2) - \Delta_3$
13:     $\ell(i, l_1, i, h_2) := \ell(i, l_1, i, h_2) + \Delta_3$
14:     **for** $\delta = 1, \ldots, \alpha_i - 1$ and all subsets A^* of $\{(i, k_1, i, k_2) \in \mathcal{A}_{\mathcal{J}} : k_2 = h_2\}$ that include
        arc (i, l_1, i, h_2) **do**
15:         **if** $\sum_{(i, h, i, h_2) \in A^*} \ell(i, h, i, h_2) > \ell(i, h_2, i, j_2) + |A^*| - 1$ **then**
16:             $\Delta_4 := (\sum_{(i, h, i, h_2) \in A^*} \ell(i, h, i, h_2)) - (\ell(i, h_2, i, j_2) + |A^*| - 1)$
17:             $\ell(i, l_1, i, h_2) := \ell(i, l_1, i, h_2) - \Delta_4$
18:         **end if**
19:     **end for**
20:     **if** $\ell(i, h_1, i, h_2) > 0$ and $\ell(i, l_1, i, l_2) > 0$ **then**
21:         **if** $\ell(i, h_1, i, h_2) < \ell(i, l_1, i, l_2)$ **then**
22:             $\ell(i, h_1, i, h_2) := 0$
23:         **else**
24:             $\ell(i, l_1, i, l_2) := 0$
25:         **end if**
26:     **end if**
27: **end if**

To delete the crossing conflict of arcs (i, h_1, i, h_2) and (i, l_1, i, l_2) the smaller value of $\ell(i, h_1, i, h_2)$ and $\ell(i, l_1, i, l_2)$ is set to zero. Reassignment of the deleted ℓ-value to $\ell(i, l_1, i, h_2)$ is tried while taking the feasibility of "Lane Consistency I" (3.10) and "Lane Consistency II" (3.13) into account.

The algorithm to correct a crossing conflict prohibited is stated in Pseudocode 6 for all arcs (i, h_1, i, h_2) and (i, l_1, i, l_2) (with $h_1 = l_2$) be arcs that cross following constraints "Prohibition of crossing conflicts II, III" (3.16) and (3.17).

If arcs (i, h_1, i, h_2) and (i, l_1, i, l_2) cause a crossing conflict (controlled in Line 1 of Pseudocode 6) then the values of number of lanes needs to be corrected. If $\ell(i, h_1, i, h_2)$ is the smaller value of $\ell(i, h_1, i, h_2)$ and $\ell(i, l_1, i, l_2)$ then value $\Delta_1 := \min\{\ell(i, h_1, i, h_2), \mathcal{L}(i, l_1, i, h_2) - \ell(i, l_1, i, h_2) - \ell(i, h_2, i, l_1), \ell(i, h_2, i, j) - \ell(i, l_1, i, h_2)\}$ is added to $\ell(i, l_1, i, h_2)$ (Line 3). Value $\mathcal{L}(i, l_1, i, h_2) - \ell(i, l_1, i, h_2) - \ell(i, h_2, i, l_1)$ is the greatest value than can be added to $\ell(i, h_1, i, l_2)$ without violating "Lane Consistency I" (3.10). Value $\ell(i, h_2, i, j) - \ell(i, l_1, i, h_2)$ is the greatest value that can be added to $\ell(i, h_1, i, l_2)$ without violating constraint

---

Pseudocode 6 Step 3.2 Repairing crossing conflicts for the case $h_1 = l_2$

1: **if** $\ell(i, h_1, i, h_2) > 0$ and $\ell(i, l_1, i, l_2) > 0$ **then**
2:   **if** $\ell(i, h_1, i, h_2) < \ell(i, l_1, i, l_2)$ **then**
3:     $\ell(i, l_1, i, h_2) := \ell(i, l_1, i, h_2) + \Delta_1$
      where $\Delta_1 := \min\{\ell(i, h_1, i, h_2), \mathcal{L}(i, l_1, i, h_2) - \ell(i, l_1, i, h_2) - \ell(i, h_2, i, l_1),$
      $\ell(i, h_2, i, j) - \ell(i, l_1, i, h_2)\}$ and where $(i, h_2, i, j) \in \mathcal{A}_C$
4:     $\ell(i, h_1, i, h_2) := 0$
5:   **else**
6:     $\ell(i, l_1, i, h_2) := \ell(i, l_1, i, h_2) + \Delta_2$
      where $\Delta_2 := \min\{\ell(i, l_1, i, l_2), \mathcal{L}(i, l_1, i, h_2) - \ell(i, l_1, i, h_2) - \ell(i, h_2, i, l_1),$
      $\ell(i, h_2, i, j) - \ell(i, l_1, i, h_2)\}$ and where $(i, h_2, i, j) \in \mathcal{A}_C$
7:     $\ell(i, l_1, i, l_2) := 0$
8:   **end if**
9:   **for** $\delta = 1, \ldots, \alpha_i - 1$ and all subsets A^* of $\{(i, k_1, i, k_2) \in \mathcal{A}_{\mathcal{J}} : k_2 = h_2\}$ that include arc (i, l_1, i, h_2) **do**
10:     **if** $\sum_{(i, h, i, h_2) \in A^*} \ell(i, h, i, h_2) > \ell(i, h_2, i, j) + |A^*| - 1$ **then**
11:       $\ell(i, l_1, i, h_2) := \ell(i, l_1, i, h_2) - \Delta_3$
      with $\Delta_3 := (\sum_{(i, h, i, h_2) \in A^*} \ell(i, h, i, h_2)) - (\ell(i, h_2, i, j) + |A^*| - 1)$
12:     **end if**
13:   **end for**
14: **end if**

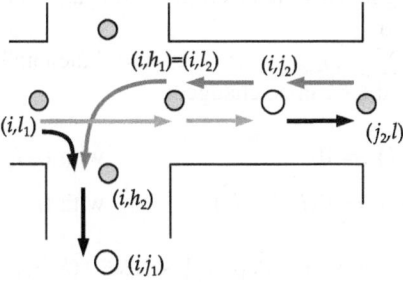

Fig. 4.12 Illustration of a solution of the relaxation of the urban evacuation model (depicted are arcs with positive number of lanes)

"Lane Consistency II" (3.13) with $\delta = 1$. At most $\ell(i, h_1, i, h_2)$ shall be added to $\ell(i, l_1, i, h_2)$, hence the value $\min\{\ell(i, h_1, i, h_2), \mathcal{L}(i, l_1, i, h_2) - \ell(i, l_1, i, h_2) - \ell(i, h_2, i, l_1), \ell(i, h_2, i, j) - \ell(i, l_1, i, h_2)\}$ is added. Then $\ell(i, h_1, i, h_2)$ is set to zero (Line 4). The reallocation of number of lanes is analog if $\ell(i, l_1, i, l_2) \leq \ell(i, h_1, i, h_2)$ (Lines 5–7). In Lines 9–13 "Lane consistency II" (3.13) is controlled for exit (i, h_2) and, if necessary, corrected.

"Deleting" the lanes of arc (i, h_1, i, h_2) or (i, l_1, i, l_2) may lead to further corrections of number of lanes as considered in Consideration 5. They may lead to similar correcting operations that follows after decreasing the number of lanes of arcs in $\mathcal{A}_C$ in Step 2 to zero. Figure 4.12 illustrates a solution of the relaxation of the urban evacuation model with a crossing conflict. Every arc depicts a direction with a positive number of lanes. In the following, the two cases $\ell(i, h_1, i, h_2) = 0$ or $\ell(i, l_1, i, l_2) = 0$ are considered.

$\ell(i, h_1, i, h_2) = 0$ If $\sum_{(i,h_1,i,h) \in \mathcal{A}_{\mathcal{J}}} \ell(i, h_1, i, h) = 0$ then no lane emerges from entrance (i, h_1) and the following values can be changed:

$$\ell(i, j, i, h_1) := 0 \qquad\qquad \text{with } (i, j, i, h_1) \in \mathcal{A}_C$$

$$\ell(i, h_1, i, j) := \mathcal{L}(i, h_1, i, j) \qquad\qquad \text{with } (i, h_1, i, j) \in \mathcal{A}_C$$

$$\ell(j, l, i, j) \;\; := 0 \qquad\qquad \text{with } (j, l, i, j) \in \mathcal{A}_C$$

$$\ell(i, j, j, l) \;\; := \mathcal{L}(i, j, j, l) \qquad\qquad \text{with } (i, j, j, l) \in \mathcal{A}_C$$

See Fig. 4.12 for an illustration. If $\ell(i, h_1, i, h_2) = 0$ (the dark gray arc within intersection i) then we can set $\ell(j_2, l, i, j_2)$ and $\ell(i, j_2, i, h_1)$ equal to zero (the number of lanes corresponding to the dark gray arcs (j_2, l, i, j_2) and (i, j_2, i, h_1) between intersection i and j in Fig. 4.12) and the number of lanes of the corresponding counter-arcs can be set to the number of available lanes. I.e. a number of lanes of an arc in $\mathcal{A}_C$ pointing to an entrance of an intersection, namely $\ell(i, j, i, h_1)$ (in Fig. 4.12 the number of lanes $\ell(i, j_2, i, h_1)$ of arc (i, j_2, i, h_1)), is decreased to zero, therefore Case 3 of Step 2 holds and the following operations that are needed there can be also used in this case: With analog steps as in "Controlling the neighboring intersection j" (on page 81) the generation of blind alley shall be avoided.

$\ell(i, l_1, i, l_2) = 0$ If $\sum_{(i,l,i,l_2) \in \mathcal{A}_{\mathcal{J}}} \ell(i, l, i, l_2) = 0$ then no lane leads to exit (i, l_2) then the following values can be changed:

$$\ell(i, l_2, i, j) := 0 \qquad\qquad \text{with } (i, l_2, i, j) \in \mathcal{A}_C$$

$$\ell(i, j, i, l_2) := \mathcal{L}(i, j, i, l_2) \qquad\qquad \text{with } (i, j, i, l_2) \in \mathcal{A}_C$$

In Fig. 4.12 the light gray arcs depict this case. $\ell(i, l_2, i, j_2)$ can be set to zero and to the counter direction (i, j_2, i, l_2) (dark gray arc) the maximal number of lanes $\mathcal{L}(i, j_2, i, l_2)$ can be assigned.

Step 3.3:

Reaching Step 3.3 all crossing conflicts are repaired and integrality remains to be assured. Therefore, every exit (i, h) of intersection i is investigated:

All values $\ell(i, a, i, h)$ with $0 < \ell(i, a, i, h) < 1$ are set to one for all $(i, a, i, h) \in \mathcal{A}_{\mathcal{J}}$ to avoid the generation of a blind alley.

If the constraints "Lane Consistency II" (3.13) are not feasible with all number of lanes of arcs pointing to exit (i, h) rounded up (i.e. $\lceil \lambda(i, a, i, h) \rceil$ instead of $\ell(i, a, i, h)$ for all $(i, a, i, h) \in \mathcal{A}_{\mathcal{J}}$) then the number of lanes with the lowest value of decimal places is rounded down. This procedure is repeated until the constraints "Lane Consistency II" (3.13) with all associated ℓ-variables rounded up remains valid.

4.2 Adjustments for the Evacuation Model with Waiting and the One-Way Urban Evacuation Model

For the adjustment heuristic, the values of the inflow decision variables x and the values of number of lanes decision variables ℓ are needed. These are achieved by solving the relaxation that regards at least the network flow constraints, the capacity constraints and the "Lane consistency constraints I and II" that restrict the assigned number of lanes with respect to the available number of lanes and restrict merging within intersections. The integrality of the decision variables of number of lanes is relaxed. The considered objective function is (3.1), but every other suitable objective could be used. If objective function (3.2) is considered, the corresponding auxiliary decision variables have to be included properly.

For the urban evacuation model with waiting an analog relaxation is solved in Stage 1 of the relaxation-based approach, where the decision variables w and the corresponding network flow constraints are considered. The integrality of ℓ-variables is relaxed and the constraints prohibiting crossing conflicts and the corresponding binary variables y are disregarded. Stage 2, the adjustment algorithm, is executed in the same way by using the values of the x- and ℓ-variables of the solution of the relaxation. In Stage 3 the urban evacuation model with waiting on arcs is solved optimally with fixed ℓ-variables achieved in Stage 2.

If the traffic routing is readjusted such that every street is a one-way street then the adjustment heuristic has less special cases. Crossing conflicts of two arcs that have a node in common prohibited by the constraints "Prohibition of crossing conflict II, III" (3.16) and (3.17) do not occur if every street section is turned into a one-way street. Step 2 can be changed and Step 3.2 concerning the crossing conflicts of type (3.16) and (3.17) can be neglected. Step 2 is adjusted such that the greater number of lanes of the appropriate arcs in $\mathcal{A}_C$ is set to the available number of lanes $\mathcal{L}$ and the lesser one is set to zero. Then every street connecting intersections is a one-way street and hence every intersection node is either an entrance or an exit. The relaxation-based approach remains structured in three stages: In Stage 1 the same relaxation used for the two-way case (with or without waiting) is solved optimally and the values of the variables of number of lanes and inflow variables of the relaxation build the basis to execute the adjustment approach that correct the ℓ-variables such that they are feasible subject to the constraints of the urban one-way evacuation model. The one-way urban evacuation model with fixed ℓ-variables achieved by the adjustment approach is solved to compute the inflow variables x and the value of the objective function.

4.3 Computational Study

Computational studies testing the relaxation-based heuristic approaches for the two-way and one-way case for the evacuation models with and without waiting were implemented. The considered objective function is (3.1), i.e. the minimization of

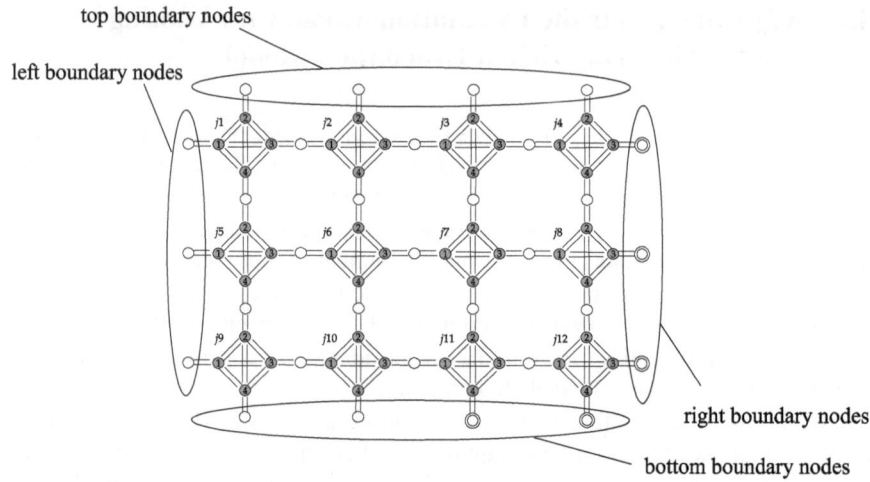

left boundary nodes

top boundary nodes

right boundary nodes

bottom boundary nodes

Fig. 4.13 Grid network of size 3×4 with exit-pattern $k = 1$

the average evacuation time can be determined. For the studies, the same set of instances were used. We built the 90 considered instances based on grid networks with intersections with $\alpha = 4$ entrances/exits. The sinks, the number of evacuees and the number of lanes are varied. Figure 4.13 shows a 3×4 grid network with sinks (nodes with a double line) on the right and the bottom boundary.

The parameters of the instances were chosen as follows:

- Five different network sizes: 3×4, 4×5, 5×5, 5×6 and 5×7
- Two different exit-patterns:

  $k = 1$: Every node is a sink at least on one side of the boundary (e.g. only on the right boundary) but at most on two sides of the boundary (in Fig. 4.13 all right boundary nodes are sinks and two nodes of the bottom boundary nodes)
  $k = 2$: Every node on the boundary is a sink

- Three variations of number of evacuees (10, 30 or 40 evacuees and a mix of 0, 10, 30 or 40 evacuees per source) each
- The number of usable lanes for all street sections was set to $\mathcal{L} = 2, 3$ or 4
- The travel time τ was chosen as follows: $\tau(j,l,i,h) = 3$ for all $(j,l,i,h) \in \mathcal{A}_\mathcal{C}$ for instances based on grid networks with exit-pattern $k = 1$ (except 3×4), $\tau(j,l,i,h) = 6$ for all $(j,l,i,h) \in \mathcal{A}_\mathcal{C}$ for instances based on grid networks with exit-pattern $k = 2$ and 3×4 with exit-pattern $k = 1$, and $\tau(i,l,i,h) = 1$ for all $(i,l,i,h) \in \mathcal{A}_\mathcal{J}$
- The inflow and total capacity were chosen as follows: the inflow capacities are $c^{in}(j,l,i,h) = 1$ for all $(j,l,i,h) \in \mathcal{A}_\mathcal{C} \cup \mathcal{A}_\mathcal{J}$, the capacities are $c(j,l,i,h) = 5$ for all $(j,l,i,h) \in \mathcal{A}_\mathcal{C}$ and $c(i,l,i,h) = 1$ for all $(i,l,i,h) \in \mathcal{A}_\mathcal{J}$

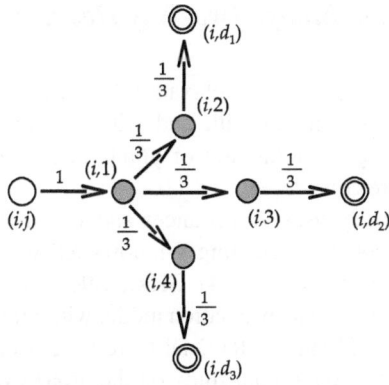

Fig. 4.14 An illustration of a solution of the relaxation of the urban evacuation model

- The considered time periods are $\mathcal{T} = \{0, 1, \ldots, 150\}$
- Parameter χ for Step 0 of the adjustment algorithm was chosen as $\chi = 0.6$

The grid network 3×4 with exit-pattern $k = 1$ is illustrated in Fig. 4.13.

For Step 0 of the adjustment approach, the value χ has to be chosen. The following problems may arise: If χ is chosen too small, it can occur that a cycle is built with no path from a certain source to a sink and an infeasible solution is generated. If χ is chosen too high, it is possible that the ℓ-variables of some street segments that are important for the feasibility of the solution are set to zero and for this reason the solution of the heuristic is infeasible. In Fig. 4.14 an example is depicted where the solution is infeasible if the value χ is chosen too high. There is a single intersection i with $\alpha_i = 4$, a source (i, j) and three sinks (i, d_1), (i, d_2) and (i, d_3). Let one evacuee be located in source (i, j). We assume that the optimal solution of the relaxation is the one depicted in Fig. 4.14. The values next to the arcs represent the values of the ℓ-variables. The number of lanes within the intersection are at most $1/3$. In this example, let this be the same as the total amount of flow on each arc. With $\chi = 0.4$ all ℓ-variables of arcs within $\mathcal{A}_{\mathcal{J}}$ and $\mathcal{A}_{\mathcal{C}}$ are set to zero (except $\ell(i, j, i, 1)$) and the resulting solution of the heuristic will be infeasible because no evacuee can reach a sink. In our computational study we choose $\chi = 0.6$ which works well for nearly all instances of the considered models. Problems arise for instances of models with waiting. A solution will be presented to avoid these problems (see the following Sect. 4.3.1) such that for all considered instances a feasible solution can be found. But it is not proved that for all instances a value for χ can be found such that a feasible solution can be found with the heuristic approach.

A solution of an instance is infeasible if the maximal number of periods T is chosen too small, here in general $T = 150$ is used. The instances based on grid networks 5×5, 5×6 and 5×7 with exit-pattern $k = 1$ and 30 evacuees in every source each are infeasible with $T = 150$. Therefore T was set to 200 in these cases.

4.3.1 Computational Study: Two-Way Heuristic Approach

For the computational study of the relaxation-based approach for the two-way case, the urban evacuation model with and without waiting with the objective function (3.1) minimizing the total sum of points in time evacuees remaining in the evacuation zone were considered.

The computational study uses 90 instances and was implemented with AMPL using CPLEX 10.0. A computer running Windows XP with an AMD Athlon(tm) 64X2 Dual Core Processor 4600+ (2.41 GHz), 1.96 GB RAM was used for the computational study on the urban evacuation model without waiting and a computer running Windows 7 with a Pentium(R) Dual-Core Processor (2.10 GHz), 4.00 GB RAM was used for the computational study on the urban evacuation problem with waiting on arcs.

The percentage gap between the result of the relaxation-based approach and the corresponding relaxation of the urban evacuation problem is computed as follows:

$$\% - \mathrm{GAP} = \frac{UB - LB}{LB} \cdot 100$$

where UB is the objective function value of the solution of the relaxation-based approach and LB is the reference value.

Computational Results Concerning the Urban Evacuation Model Without Waiting

The results are taken from Bretschneider and Kimms (2011). The reference value for instances of the evacuation model without waiting is the best lower bound or optimal solution value that was possible to compute within a given time limit. Therefore the best known lower bound for the respective instance was chosen among the following solutions:

The solution of the urban evacuation model where the computation time for instances based on grid networks 3×4, 4×5 and 5×5 with exit-patterns $k = 1, 2$ was limited to ca. 2 h (7, 300 s) and for instances based on networks 5×6 and 5×7 with exit-patterns $k = 1, 2$ it was limited to ca. 3.6 h (13, 000 s). Here it was only possible to compute a feasible or optimal solution in 40% of the instances. Or a solution of the urban evacuation model with at most a relative tolerance of 0.1 with a restriction on computation time of at most 12 h (43, 200 s) is considered. Here it was possible to compute a feasible or optimal solution in 66% of the instances. Or the solution of the relaxation of the urban evacuation model needed for the relaxation-based heuristic. For every instance the optimal solution of the relaxation was found.

Note, the reference values are not of the same quality. We refer to the Appendix A where for every instance the corresponding kind of solution and percentage gap are listed.

Table 4.1 Average percentage gaps between the results of the relaxation-based approach and the best known lower bound

Size	Exit-pattern k	Avg %-GAP
3×4	1	2.65
3×4	2	3.03
4×5	1	10.12
4×5	2	3.37
5×5	1	4.61
5×5	2	3.17
5×6	1	3.60
5×6	2	3.19
5×7	1	7.99
5×7	2	2.93

The resulting percentage gaps were aggregated over a fixed network size and a fixed exit-pattern. The resulting average gaps are shown in Table 4.1.

The average percentage gaps between the compared values vary from 2.65% to 10.12% in size. The largest percentage gap occurs for the instance based on network 5×7 with $k = 1$, $\mathcal{L} = 4$ and a mixed number of evacuees; it is 39.72%. But this is an exceptional case. 52% of the percentage gaps are less than or equal to 2%, 84% are less than or equal to 10%. And just 3% of the percentage gaps are greater than 20%.

The computation times were aggregated over a fixed network size and a fixed exit-pattern. Table 4.2 shows the average computation times that are needed to compute the results in Table 4.1. The time needed to execute the stages of the adjustment algorithm and the computation of the urban evacuation model with fixed ℓ-values are done in seconds in case of the smaller instances and at most in ca. 160 s in case of the instances of the network 5×7. The computation time of the relaxation-based heuristic approach consists of the solution time of the relaxation of the urban evacuation model, the execution of the adjustment algorithm and the solution time of the urban evacuation model with fixed ℓ-values achieved by the first two stages of the approach. The dominant time is the time needed to compute a solution of the relaxation.

The maximal solution time of the relaxation-based approach is 13, 811 s (ca. 3.8 h) for the problem instance based on grid network 5×7 with $k = 2$, $\mathcal{L} = 3$ and the number of evacuees in each source are 30 and 37, 497 s (ca. 10.4 h) for the problem instance based on grid network 5×7 with $k = 1$, $\mathcal{L} = 2$ and the number of evacuees in each source are 30. Both are instances with the increased number of time units ($T = 200$). As reference value the solution of the relaxation has been used because no other value was available. The computation times for a better lower bound (than the solution of the relaxation) with the help of the urban evacuation model are in the majority much higher than the solution times of the relaxation. For example to compute a feasible solution of the instance based on grid network 4×5, where $k = 2$, $\mathcal{L} = 2$ and the number of evacuees in each source are 40, the heuristic approach needs 176 s (ca. 3 min) and to compute the reference value 11, 189 s (ca. 190 min) are needed.

Table 4.2 Average run-times of the computation of a solution of the relaxation-based approach and the computation of the best known lower bound in CPU-seconds

Size	Exit-pattern k	Avg run-time relax.-based appr.	Avg run-time comp. of LB
3×4	1	74	25,059
3×4	2	57	351
4×5	1	442	10,029
4×5	2	188	8,445
5×5	1	3,183	11,728
5×5	2	364	8,923
5×6	1	6,421	10,620
5×6	2	439	9,828
5×7	1	7,747	11,518
5×7	2	587	1,750

The computational study was done on a slightly different representation of the network. Superfluous arcs were neglected in Sect. 3.1 that are not needed to represent the network properly. See Appendix A.2 for an example of the used network representation. It has no effects on the relaxation-based approach. That network representation are also presented in Bretschneider and Kimms (2011).

Computational Results Concerning the Evacuation Model with Waiting

The results of the computational study for the relaxation-based approach used to solve the urban evacuation model with waiting shows a drawback of the approach: finding a value of χ in Step 0 of the adjustment approach, such the generated traffic routing is a feasible solution of the evacuation problem, turns out to take a little bit more effort.

The relaxation solved in Stage 1 of the relaxation-based approach relaxes the traffic routing constraints such that no crossing conflict has to be regarded and integrality is neglected. In Stage 2, the χ-value was chosen as before as $\chi = 0.6$. To evaluate the result of the relaxation-based heuristic approach, the reference values were chosen to be the solution of the relaxation.

The resulting average percentage gaps that were aggregated over a fixed network size and exit-pattern are shown in Table 4.3 on the left. Difficulties to choose a χ-value arose in computing the heuristic approach for the instances based on networks 5×5, 5×6 and 5×7 with exit-pattern $k = 1$, $\mathcal{L} = 2, 3, 4$ and an initial number of evacuees of 30 per source. With $\chi = 0.6$ and the time horizon $\mathcal{T} = \{0, 1, \ldots, 150\}$, i.e. $T = 150$, no feasible solution could be generated, i.e. the time horizon $\mathcal{T}$ was too small or there exited no path from at least one source to a sink. In a first attempt, the relaxation-based approach was computed with different

Table 4.3 Average percentage gaps between the results of the relaxation-based approach and the relaxation and average run-times (in CPU-seconds) of the computation of a solution of the relaxation-based approach solving the evacuation model with waiting

Size	Exit-pattern k	Avg %-GAP	Size	Exit-pattern k	Avg run-time relax.-based appr.
3×4	1	3.91	3×4	1	171
3×4	2	2.73	3×4	2	122
4×5	1	11.00	4×5	1	459
4×5	2	3.16	4×5	2	359
5×5	1	9.42	5×5	1	849
5×5	2	3.51	5×5	2	545
5×6	1	9.43	5×6	1	1,262
5×6	2	3.75	5×6	2	831
5×7	1	7.08	5×7	1	1,432
5×7	2	3.10	5×7	2	1,017

$\chi \in \{0, 0.5, 1, 1.5, \ldots, 16\}$ and an increased time horizon $\mathcal{T} = \{0, 1, \ldots, 250\}$ (the length of the time periods remained the same, but the time horizon was enlarged): For instances with the available number of lanes $\mathcal{L} = 3, 4$, it was possible to generate a feasible solution.

For the instances with $\mathcal{L} = 2$ we tried two different approaches: first, we increased the time horizon, consecutively, to $T = 350, 450, 550, 650, 750$ and second, we considered the modification choosing different χ-values for different arcs. Both variants of the approach could generate feasible solutions for some T- and χ-values and the best solutions were considered in Table 4.3. These are the feasible solutions generated with the second variant.

The average percentage gaps between the compared values vary from 2.73% to 11.00% in size. The largest percentage gap occurs for instance based on network 5×6 with $k = 1$, $\mathcal{L} = 2$ and a number of evacuees of 10 per source; it is 33.73%. Similar as to that which was shown with the computational study solving the evacuation model without waiting, this is an exceptional case: 50.57% of the percentage gaps of the considered instances are less than or equal to 2%, 79.31% are less than or equal to 10% and just 3.45% of the percentage gaps are greater than 20%.

The computation times were aggregated over a fixed network size and a fixed exit-pattern. Table 4.3 on the right shows the average computation times that are needed to compute the results in Table 4.3 on the right.

The time needed to execute the stages of the adjustment algorithm is completed in a range of 7–640 s, the latter computation time is needed for the instance based on network 5×7, $k = 1$, $\mathcal{L} = 2$ and 30 evacuees per source. The computation of the urban evacuation model with waiting with fixed ℓ-values are completed in around 60 s on average and at most in ca. 400 s in case of the instances of the network 5×7. The computation time of the relaxation lies between 79 s and 1, 582 s (i.e. ca. 1 min and ca. 26 min).

Table 4.4 Percentage gaps of solutions of the relaxation-based approach solving the evacuation model with waiting of certain instances

Size	Exit-pattern k	Avg %-GAP(T)	Avg %-GAP(χ)	Avg %-GAP(objective)
5×5	1	59.35	0.00	0.00
5×6	1	76.11	0.00	0.00
5×7	1	92.82	0.00	0.01

The results of the two different approaches to achieve a feasible solution with the relaxation-based heuristic of the instances based on the networks 5×5, 5×6 and 5×7 with exit-pattern $k = 1$, available number of lanes $\mathcal{L} = 2$ and 30 evacuees in each source are the following: For the instance based on network 5×5, $\chi = 0.6$ and an increased number of time units $T = 650$ a feasible solution could be computed. For the two instances based on networks 5×6 and 5×7, $\chi = 0.6$ and an increased time horizon $T = 750$, feasible solutions could be generated, too. I.e. all flow units can reach the sinks within the time horizon $\mathcal{T} = \{0, \ldots, T\}$.

For the second variant, the value of χ was set to 10 for all arcs of intersections next to a dead end source because the flow that arises from the source at the boundary are 30 flow units. They can be led to three different directions within the neighboring intersection. For the rest of the arcs, $\chi \in \{20, 22, \ldots, 30\}$ were tested and with $\chi = 28$ the best results are achieved. In Table 4.4 the results of increasing the time horizon, indicated with T, and the approach of choosing different values for χ in Step 0 of the relaxation-based approach, indicated with χ, are shown. The results for the second variant are the optimal solutions each. A third possibility will be presented next; the solutions are indicated with *objective*.

The problems that lead to the infeasibility arise from flows that circle in the network "waiting" indirectly that capacity is free to travel to a sink. Even if waiting on arcs is allowed, it is not prohibited that flow moves within the network instead of waiting on an arc. The objective function just requires that the average time the evacuees are within the evacuation zone is minimized, but it does not matter where the evacuees are in the network. Hence, flow can cycle until capacity towards a sink is free. It is possible that some pathways even in the optimal solution are not necessary or may be traveled multiple times. For an illustration see Fig. 4.15. The network is depicted partly. Intersections j_1 to j_{15} and the sink nodes are shown. The sinks of the network are depicted again as cycles with a double line. A solution of the relaxation-based heuristic approach of the instance based on network 5×5 with exit pattern $k = 1$, available number of lanes $\mathcal{L} = 2$ and 30 evacuees in every source is depicted. The cycling of flow is visible in the values of number of lanes. Here, as well it becomes apparent why the time horizon has to be increased to achieve a feasible solution. The only exit of sources located above the intersections j_5, j_6, ... j_{10} is through intersection j_5. Therefore the time needed such that all evacuees can enter a sink has to be increased extremely.

One possibility to improve the heuristic approach would be to investigate the choice of the value χ. We decided to choose another way and to adjust the objective of the relaxation. To avoid unnecessary ways (similar to the approach of Hamacher

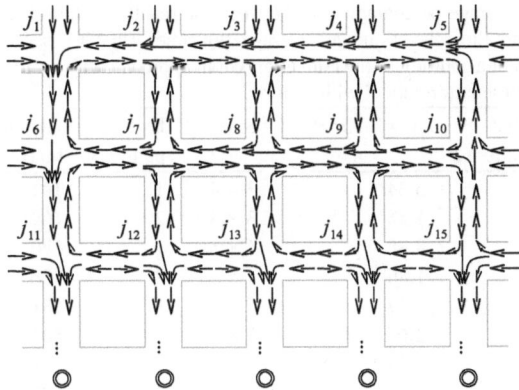

Fig. 4.15 An illustration of a solution of the relaxation-based heuristic for an instance based on network 5 × 5

and Tufekci 1987) and to encourage the flow to wait, the objective function of the relaxation of the urban evacuation model with waiting is extended to

$$\sum_{(d_1,d_2)\in\mathcal{D}}\sum_{t\in\mathcal{T}}t\cdot x(d_1,d_2,D,0,t) + \sum_{(j,l,i,h)\in\mathcal{A}_C\cup\mathcal{A}_{\mathcal{J}}}\sum_{t\in\mathcal{T}}\tau(j,l,i,h)\cdot x(j,l,i,h,t).$$

The second term leads to penalty if inflow occurs. Hence, waiting on arcs is not penalized, but moving within the street network is penalized. The penalty costs are set to the travel time and are not dependent on time. If the cost on arcs was dependent on time then unnecessary ways may lead to a better objective value, but the intention to extend the objective function of the relaxation was because that should be avoided.

Changing the objective function of the relaxation of the urban evacuation model and execute the relaxation-based heuristic approach as before leads to results depicted in Table 4.5. For every instance, a feasible solution is generated with the relaxation-based approach using the value $\chi = 0.6$.

The average percentage gaps are shown in Table 4.5 on the left and the average computation times are shown in Table 4.5 on the right. The largest percentage gap occurs for instance based on network 5 × 7 with $k = 1$, $\mathcal{L} = 4$ and a mixed number of evacuees; the percentage gap is 27.16%. 61.11% of the percentage gaps of the considered instances are less than or equal to 2%, 85.56% are less than or equal to 10% and just 3.33% of the percentage gaps are greater than 20%.

The computation time of the adjustment algorithm is between 6 to 180 s and the computation of the urban evacuation model with waiting and with fixed ℓ-values are done between 2 and 80 s, on average in 15 s. The computation time of the relaxation lies between 63 s and 1,086 s (i.e. ca. 1 min and ca. 18 min).

The result of the number of lanes computed with the varied relaxation-based approach of the considered instance above, i.e. the instance based on network 5 × 5 with exit pattern $k = 1$, available number of lanes $\mathcal{L} = 2$ and 30 evacuees in every

Table 4.5 Average percentage gaps between the results of the relaxation-based approach and the relaxation and average run-times (in CPU-seconds) of the computation of a solution of the relaxation-based approach with a varied objective solving the evacuation model with waiting

Size	Exit-pattern k	Avg %-GAP	Size	Exit-pattern k	Avg run-time relax.-based appr.
3×4	1	3.34	3×4	1	141
3×4	2	1.58	3×4	2	109
4×5	1	8.88	4×5	1	370
4×5	2	1.86	4×5	2	336
5×5	1	4.67	5×5	1	581
5×5	2	2.65	5×5	2	504
5×6	1	5.59	5×6	1	751
5×6	2	2.19	5×6	2	693
5×7	1	5.77	5×7	1	908
5×7	2	1.59	5×7	2	938

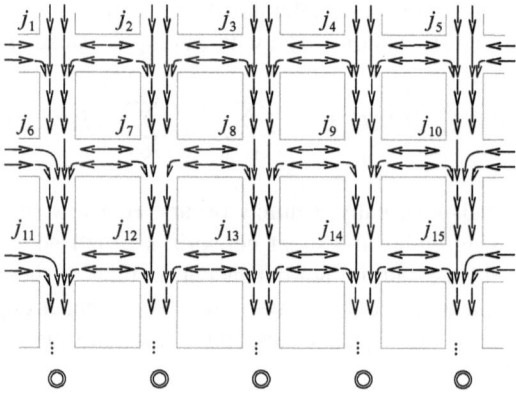

Fig. 4.16 An illustration of a solution of the relaxation-based heuristic with the variation of the objective of the relaxation for an instance based on network 5×5

source, is illustrated partly in Fig. 4.16. In this case, no cycling occurs and just one-way streets are generated.

For all the considered instances, it is always possible to generate a feasible traffic routing such that all evacuees can reach a safe zone within the given time horizon ($T = 150$ and $T = 200$ for the three above-mentioned instances based on networks 5×5, 5×6 and 5×7). But an open question is whether this is possible for every instance.

4.3.2 Computational Study: One-Way Heuristic Approach

The urban evacuation models with and without waiting optimizing the objective function (3.1) are considered for the computational study of the relaxation-based approach for the one-way case. The computational study used the same 90 instances as above and was implemented with AMPL using CPLEX 10.0. A computer running Windows 7 with a Pentium(R) Dual-Core Processor (2.10 GHz), 4.00 GB RAM was used for the computational study on the urban one-way evacuation problem without and with waiting on arcs.

The percentage gap between the result of the relaxation-based approach and the relaxation of the urban evacuation problem is computed as follows:

$$\% - \text{GAP} = \frac{UB - LB}{LB} \cdot 100$$

where UB is the objective function value of solution of the relaxation-based approach, and LB is the value of the weighted sum of flows of arcs pointing to the super sink (i.e. the value of the objective function (3.1)). I.e. the objective function value of the relaxation considered in Stage 1 of the relaxation-based heuristic for the model without waiting. For the model with waiting on arcs, the objective function of the relaxation of Stage 1 is the weighted sum of flows of arcs pointing to the super sink and additional the sum of penalty costs for all arcs of the street network (i.e. arcs in $\mathcal{A}_C \cup \mathcal{A}_\mathcal{J}$) for every point in time. For the computation of the percentage gap $\% - GAP$, only the value of the weighted sum of flows of arcs to the super sink, i.e. the value of the objective (3.1), is considered.

The results executing the relaxation-based heuristic approach for the one-way case on the instances are depicted for the urban evacuation model without waiting in Table 4.6 and for the urban evacuation model with waiting in Table 4.8. The relaxation for the heuristic for the model with waiting is extended with penalty costs on arcs.

Results of the One-Way Model Without Waiting Computed
with the Relaxation-Based Heuristic

The average percentage gaps are shown in Table 4.6 on the left and the average computation times are shown in Table 4.6 on the right. The largest percentage gap occurs for the instance based on network 5×7 with $k = 1$, $\mathcal{L} = 4$ and a mixed number of evacuees; the percentage gap is 43.55%. Compared to the result of the two-way heuristic (that is 39.72%), this value is higher. 50.00% of the percentage gaps of the considered instances are less than or equal to 2%, 82.22% are less than or equal to 10% and just 4.44% of the percentage gaps are greater than 20%.

The time needed to operate the adjustment algorithm on the instances lies between 3 to 63 s and the computation of the urban evacuation model with fixed ℓ-values are done between 0.5 and 14 s. The computation time of the relaxation lies between ca. 0.5 min and ca. 8 h, it is the same relaxation used for the solution of the heuristic approach in the two-way case (but computed on another computer again).

Compared to the results of the two-way relaxation-based heuristic approach, the values of the objective functions are similar. On the average they differ in 0.26%.

Table 4.6 Average percentage gaps between the results of the relaxation-based approach and the relaxation and average run-times (in CPU-seconds) of the computation of a solution of the relaxation-based approach solving the one-way evacuation model

Size	Exit-pattern k	Avg %-GAP	Size	Exit-pattern k	Avg run-time relax.-based appr.
3×4	1	4.53	3×4	1	51
3×4	2	3.17	3×4	2	40
4×5	1	10.97	4×5	1	323
4×5	2	3.53	4×5	2	127
5×5	1	4.11	5×5	1	2,633
5×5	2	3.59	5×5	2	240
5×6	1	4.11	5×6	1	5,683
5×6	2	3.54	5×6	2	299
5×7	1	8.98	5×7	1	6,958
5×7	2	2.87	5×7	2	394

Table 4.7 Average percentage gaps between the objective function values of the relaxation-based approach for the one-way and two-way evacuation model without waiting

Size	Exit-pattern k	Avg %-GAP$_H$
3×4	1	1.10
3×4	2	0.06
4×5	1	0.79
4×5	2	0.07
5×5	1	−0.43
5×5	2	0.10
5×6	1	0.40
5×6	2	0.19
5×7	1	0.74
5×7	2	−0.08

The considered percentage gap to compare the objective function values of the results of the one-way, denoted with H_{one}, and two-way approach, denoted with H_{two}, is computed with:

$$\% - \text{GAP}_H = \frac{H_{one} - H_{two}}{H_{two}} \cdot 100$$

The average percentage gaps $\% - \text{GAP}_H$ aggregated over a fixed network size and a fixed exit-pattern are presented in Table 4.7. The average percentage gaps are very small and vary from −0.43% to 1.10% on average. In 14.44% of the considered instances, the objective value generated with the one-way heuristic approach is better than the objective value of the two-way heuristic approach, in 36.67% of the considered instances a better solution could be generated with the two-way heuristic approach and in 48.89% of the instances the objective values are the same.

Table 4.8 Average percentage gaps between the results of the relaxation-based approach and the relaxation and average run-times (in CPU-seconds) for obtaining a solution for the one way evacuation model with waiting using the relaxation-based approach with a varied objective

Size	Exit-pattern k	Avg %-GAP	Size	Exit-pattern k	Avg run-time relax.-based appr.
3×4	1	3.34	3×4	1	142
3×4	2	1.58	3×4	2	110
4×5	1	8.56	4×5	1	359
4×5	2	1.86	4×5	2	331
5×5	1	4.67	5×5	1	593
5×5	2	2.65	5×5	2	498
5×6	1	5.78	5×6	1	773
5×6	2	2.19	5×6	2	694
5×7	1	5.64	5×7	1	936
5×7	2	1.59	5×7	2	940

Results of the One-Way Model with Waiting on Arcs Computed with the Relaxation-Based Heuristic

The one-way urban evacuation model is also considered with the underlying network flow model with waiting on arcs. The results are shown in Table 4.8. On the left, the average percentage gaps are depicted and on the right, the average computation times are shown.

The largest percentage gap occurs for the instance based on network 5×7 with $k = 1$, $\mathcal{L} = 4$ and a mixed number of evacuees; the percentage gap is 26.92%. Compared to the worst case result of the two-way heuristic with waiting (that is 27.16%), this value is slightly smaller. Considering the percentage gaps of each instance the following can be observed: 61.11% of the percentage gaps of the investigated instances are less than or equal to 2% (the same result occurs in the two-way case), 87.78% are less than or equal to 10% and just 3.33% of the percentage gaps are greater than 20%.

The time needed to operate the adjustment algorithm on the considered instances lies between 8 to 136 s and the computation of the urban evacuation model with waiting with fixed ℓ-values are done between 2 and 130 s. The computation time of the relaxation lies between ca. 1 min and ca. 26 min, it is the same relaxation used for the solution of the heuristic approach in the two-way case with waiting.

Comparing the objective values of the one-way and two-way approach with waiting, it can be observed that the results for most of the instances are the same, in 88.89% of the considered instances the same objective value is computed with the corresponding heuristic approach. In 5.56% of the considered instances a better objective function is achieved with the two-way approach and in 5.56% of the considered instances a better objective function is achieved with the one-way approach. The objective values are compared with the percentage gap GAP_H, presented above.

References

Bretschneider S, Kimms A (2011) A basic mathematical model for evacuation problems in urban areas. Transport Res A Pol Pract 45:523–539

Hamacher HW, Tufekci S (1987) On the use of lexicographic min cost flows in evacuation modeling. Nav Res Logist 34:487–503

Chapter 5
A Pattern-Based Evacuation Planning Model for Urban Areas

The urban evacuation model that will be presented in this chapter is a pattern-based dynamic network flow model without waiting. It is based on a detailed network, like the model presented in Chap. 3, that models every entrance/exit of every intersection and every possible direction within these intersections to map the traffic conflicts within them. The problem stays the same as in Chap. 3. The traffic routing for an emergency evacuation is to be determined such that the flow of evacuees traveling from sources to sinks is optimized. The considered objective in this chapter is the weighted sum of flows entering the sinks, where the weights are increasing with time. We assume that every person can leave the evacuation zone on its own by a vehicle. To provide a safe evacuation the traffic routing is adjusted subject to the following traffic management ideas: the readjustment of every street to a one-way street, whereas a street can be divided into two sections and these sections can lead in opposing directions, the prohibition of crossing conflicts and the limitation of the number of merging lanes within intersections.

For the optimization model every feasible combination of numbers of lanes is constructed in advance and is united in patterns for every intersection and for every street between two intersections afterwards. Every pattern determines the inflow and total flow capacities of the arcs associated with these patterns. Exactly one pattern for every intersection and for every street-connection has to be chosen. That leads to an evacuation model that is a pattern-based discrete dynamic network flow model.

In Sect. 5.1, the pattern-based urban evacuation model will be presented (compare Bretschneider and Kimms, 2011). In Sects. 5.2 and 5.3 patterns of street sections between intersections and patterns of intersections are investigated, respectively.

5.1 A One-Stage Pattern-Based Urban Evacuation Model

Like for the construction of the network in Sect. 3.1 the set $\mathcal{J}$ of intersections, the subset $\mathcal{J}_D \subset \mathcal{J}$ of intersections that are defined as destinations and the set $\mathcal{P}$ of all pairs of intersections that are directly connected are needed to generate the detailed

S. Bretschneider, *Mathematical Models for Evacuation Planning in Urban Areas*,
Lecture Notes in Economics and Mathematical Systems 659,
DOI 10.1007/978-3-642-28759-6_5, © Springer-Verlag Berlin Heidelberg 2013

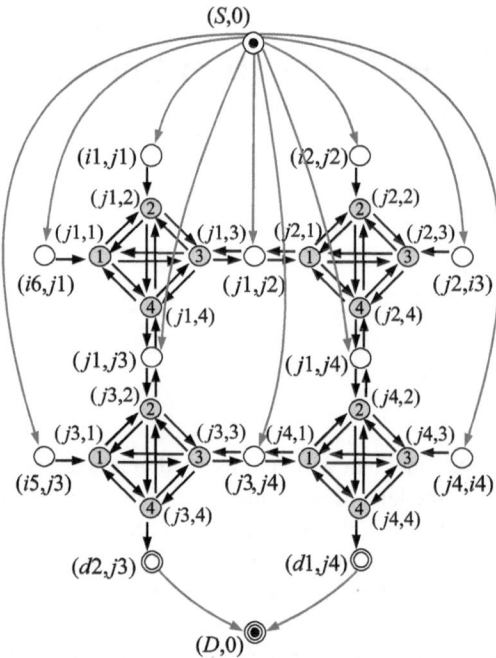

Fig. 5.1 An illustration of the nodes and the arcs in the network **G**

network. The sets of sources $\mathcal{S}$, sinks $\mathcal{D}$, nodes $\mathcal{N}_{\mathcal{J}}$ and arcs $\mathcal{A}_{\mathcal{J}}$ of intersections are generated the same way. Note that $(i, j) \in \mathcal{P}$ and $(j, i) \in \mathcal{P}$ stand for the same node and $\mathcal{P} = \mathcal{S} \cup \mathcal{D}$.

However, the set $\mathcal{A}_C$ of arcs is generated a little bit differently. The arcs pointing from an intersection to a blind alley are directly omitted: $\mathcal{A}_C$ is the set of arcs that connect sources and sinks $(i, j) \in \mathcal{S} \cup \mathcal{D}$ with the appropriate entrance/exit nodes of the intersections i and j. The arcs pointing in both directions are included, if the number of entrances/exits of i and j is greater than one (i.e. $\alpha_i > 1$ and $\alpha_j > 1$), thus the arcs (i, j, i, l), (i, l, i, j), (i, j, j, h) and (j, h, i, j) are included where (i, l) and (j, h) are the appropriate nodes of intersections. If $(i, j) \in \mathcal{S}$ and $\alpha_i = 1$ then the arc pointing from (i, j) to the appropriate entrance node (j, h) of the intersection j is included in set $\mathcal{A}_C$, i.e. arc (i, j, j, h). If $(i, j) \in \mathcal{D}$ with $\alpha_i = 1$ then just the arc pointing from the appropriate exit node (j, h) of intersection j to the sink node (i, j) belongs to $\mathcal{A}_C$, i.e. arc (j, h, i, j). A super source $(S, 0)$, a super sink $(D, 0)$ and the associated sets of arcs $\mathcal{A}_S$ and $\mathcal{A}_D$ are added as well. The complete street network is represented by the network $\mathbf{G} = (\mathcal{N}, \mathcal{A})$ with $\mathcal{N} = \{(S, 0)\} \cup \mathcal{S} \cup \mathcal{N}_{\mathcal{J}} \cup \mathcal{D} \cup \{(D, 0)\}$ and $\mathcal{A} = \mathcal{A}_S \cup \mathcal{A}_C \cup \mathcal{A}_{\mathcal{J}} \cup \mathcal{A}_D$.

Figure 5.1 illustrates the network $\mathbf{G} = (\mathcal{N}, \mathcal{A})$ of a street network with 12 intersections, where two of the intersections with one exit are destinations. It shows four intersections ($j1, j2, j3$ and $j4$) with four entrances/exits each and eight intersections ($i1, \ldots, i6, d1, d2$) with one entrance or exit with the associated sources each, super nodes and the corresponding arcs. There are six dead end sources $(i1, j1)$, $(i2, j2)$, $(j2, i3)$, $(j4, i4)$, $(i5, j3)$, and $(i6, j1)$ and two sink nodes $(d2, j3)$ and $(d1, j4)$.

Patterns are generated for arcs within intersections and for arcs corresponding to street sections between two intersections. We will consider non-dominated patterns to solve the pattern-based urban evacuation model. A pattern is non-dominated if it is not included in any other pattern or, in other words, if it is not possible to add one lane without violating a constraint. We will introduce patterns in more detail in Sects. 5.2 and 5.3.

A pattern for intersection i contains the number of lanes for every arc within i, i.e. the arcs (i, h, i, l) with $h, l \in \{1, \cdots, \alpha_i\}$ and $h \neq l$. For every $i \in \mathcal{J}$, $G_{\mathcal{J}}$ types of patterns are distinguished. These types $type(i) \in \{1, \ldots, G_{\mathcal{J}}\}$, $i \in \mathcal{J}$, differ in the number of usable lanes, the number of entrances/exits of the intersection, the allowed velocity and the length of the associated street segment which influences the parameters inflow and total capacity that have to be determined with patterns. A pattern of an intersection is feasible if the number of usable lanes is not exceeded, the one-way constraint is valid, no crossing of two directions (represented by arcs) is allowed and the merging of directions is restricted properly. Let $\mathcal{PJ}$ be the set of all feasible patterns of intersections.

Let $\mathcal{PC}$ be the set of all feasible patterns associated with street sections between two intersections. Consider two intersections i and j, where $(i, j) \in \mathcal{S}$ and $\alpha_i > 1$, $\alpha_j > 1$, then a pattern includes the number of lanes of the arcs (i, l, i, j), (i, j, i, l), (j, h, i, j) and (i, j, j, h), where (i, l) and (j, h) are the appropriate nodes of intersections i and j. If $(i, j) \in \mathcal{S}$ where one intersection has just one entrance/exit, assume $\alpha_i = 1$, then a pattern includes the number of lanes of the arc (i, j, j, h) with (j, h) as the appropriate entrance node of intersection j. If $(i, j) \in \mathcal{D}$ is considered, where $\alpha_i = 1$, then a pattern includes the number of lanes of the arc (j, h, i, j) (with (j, h) as the appropriate entrance node). For every $(i, j) \in \mathcal{P}(= \mathcal{S} \cup \mathcal{D})$, G_C different types of patterns are distinguished. These types $type(i, j) \in \{1, \ldots, G_C\}$, $(i, j) \in \mathcal{P}$, denote the number of usable lanes, the allowed velocity and the length of the associated street segment (i, j). The parameters assigned to a pattern (inflow and total capacity) are different if the numbers of lanes, the allowed velocity or the length of the street segment, i.e. the pattern-types, differ. A pattern of $(i, j) \in \mathcal{P}$ is feasible if the number of usable lanes is not exceeded and if on every street section a positive number of lanes is assigned to only one direction, i.e. every street section is a one-way street.

The sets of patterns $\mathcal{PJ}$ and $\mathcal{PC}$ have to be generated before the optimization.

The time horizon is subdivided into T time units, hence we consider the set of $T + 1$ points in time $\mathcal{T} = \{0, \ldots, T\}$.

We need the following parameters and sets to formulate the pattern-based optimization model:

- $\mathcal{PC} = \bigcup_{g=1}^{G_C} \mathcal{PC}(g)$ is the set of patterns for the arcs that connect two intersections
- $\mathcal{PJ} = \bigcup_{g=1}^{G_{\mathcal{J}}} \mathcal{PJ}(g)$ is the set of patterns for the arcs within intersections
- $O(i, j)$ is the number of evacuees that are assigned to source $(i, j) \in \mathcal{S}$

- $\tau(j,l,i,h) \in \mathbb{N}$ is the travel time on arc $(j,l,i,h) \in \mathcal{A}$, $\tau(j,l,i,h) = 0$ for all $(j,l,i,h) \in \mathcal{A}_\mathcal{S} \cup \mathcal{A}_\mathcal{D}$
- $c^{in}(j,l,i,h,p)$ is the inflow capacity of arc $(j,l,i,h) \in \mathcal{A}_C$ in pattern $p \in \mathcal{PC}$
- $c(j,l,i,h,p)$ is the total capacity of arc $(j,l,i,h) \in \mathcal{A}_C$ in pattern $p \in \mathcal{PC}$

The decision variables are defined as follows:

- $x(j,l,i,h,t) \in \mathbb{R}_0^+$ is the inflow of vehicles which leave node (j,l) in time $t \in \mathcal{T}$ to reach node (i,h) in time $t + \tau(j,l,i,h)$ with $(j,l,i,h) \in \mathcal{A}, t \in \mathcal{T}$. For the ease of notation we define $x(j,l,i,h,t) = 0$ for $t < 0$.
- $y_C(i,j,p) \in \{0,1\}$, where $y_C(i,j,p) = 1$ if pattern $p \in \mathcal{PC}$ is chosen for the connection $(i,j) \in \mathcal{P}$, $y_C(i,j,p) = 0$ otherwise.
- $y_\mathcal{J}(j,p) \in \{0,1\}$, where $y_\mathcal{J}(j,p) = 1$ if pattern $p \in \mathcal{PJ}$ is chosen for intersection $j \in \mathcal{J}$, $y_\mathcal{J}(j,p) = 0$ otherwise.

The pattern-based evacuation model can be sub-divided into the parts objective function, network flow, traffic routing and capacity constraints.

Objective Function

The weighted sum of flows entering the arcs that lead to the super sink is to be minimized, where the function $\gamma : \mathcal{T} \rightarrow \mathbb{R}$ is increasing with time.

$$\min \sum_{t \in \mathcal{T}} \sum_{(d,i) \in \mathcal{N}_\mathcal{D}} \gamma(t) \cdot x(d,i,D,0) \tag{5.1}$$

Network Flow Constraints

The network flow constraints are a subset of the network flow constraints of the urban evacuation problem presented in Chap. 3. We will repeat them shortly. The constraints to prohibit irregular flow are not needed if the patterns represent one-way streets. We will consider a traffic routing for the case of an evacuation that only consists of one-way streets and the available choice of traffic routing patterns have to take that into account.

- Flow enforcement:
  All vehicles of evacuees have to leave the corresponding sources within the time horizon and all vehicles have to enter a sink within the considered time horizon.

$$\sum_{t \in \mathcal{T}} x(S,0,s_1,s_2,t) = O(s_1,s_2) \text{ for all } (s_1,s_2) \in \mathcal{S} \tag{5.2}$$

$$\sum_{t \in \mathcal{T}} \sum_{(d_1,d_2) \in \mathcal{D}} x(d_1,d_2,D,0,t) = \sum_{(s_1,s_2) \in \mathcal{S}} O(s_1,s_2) \tag{5.3}$$

- Flow conservation constraint:

$$\sum_{(i,h)\in\mathcal{N}:(i,h,j,l)\in\mathcal{A}} x(i,h,j,l,t-\tau(i,h,j,l)) = \sum_{(i,h)\in\mathcal{N}:(j,l,i,h)\in\mathcal{A}} x(j,l,i,h,t)$$

(5.4)

$$\text{for all } (j,l) \in \mathcal{N}\backslash\{(S,0),(D,0)\}; t \in \mathcal{T}$$

Traffic Routing Constraints

All restrictions on the traffic routing are included implicitly in the patterns. These patterns have to be determined before an instance of the pattern-based urban evacuation model can be solved. Patterns determine the parameters inflow and total capacity of the corresponding street sections.

- Exactly one pattern has to be chosen for every intersection $i \in \mathcal{J}$ with $\alpha_i \geq 2$. These patterns determine the parameters of every street section, i.e. every possible turn within an intersection.

$$\sum_{p\in\mathcal{P}\mathcal{J}(type(i))} y_{\mathcal{J}}(i,p) = 1 \quad \text{for all } i \in \mathcal{J} : \alpha_i \geq 2 \qquad (5.5)$$

- Exactly one pattern has to be chosen for every street section that connects two intersections including the street segments leading to a sink or corresponding to a blind alley, i.e. for every $(i,j) \in \mathcal{P}$.

$$\sum_{p\in\mathcal{P}\mathcal{C}(type(i,j))} y_C(i,j,p) = 1 \text{ for all } (i,j) \in \mathcal{P} \qquad (5.6)$$

The patterns may be of different type $type(i)$ and $type(k,j)$ for intersections i and street segments (k,j), respectively. These types correspond to the number of usable lanes, the length of a street, the allowed velocity and, in case of the intersections, to the number of entrances/exits.

Capacity Constraints

- Inflow capacity constraints:
  The restriction of the inflow for every point in time depends on the chosen pattern for every street segment within an intersection (with $\alpha \geq 2$) and for every street between two intersections.

$$x(i,l,i,h,t) \leq \sum_{p \in \mathcal{PJ}(type(i))} c^{in}(i,l,i,h,p) y_{\mathcal{J}}(i,p) \qquad (5.7)$$

$$\text{for all } (i,l,i,h) \in \mathcal{A}_{\mathcal{J}}; t \in \mathcal{T}$$

$$x(i,l,i,j,t) \leq \sum_{p \in \mathcal{PC}(type(i,j))} c^{in}(i,l,i,j,p) y_C(i,j,p) \qquad (5.8)$$

$$\text{for all } (i,l,i,j) \in \mathcal{A}_C \text{ with } (i,l) \in \mathcal{N}_{\mathcal{J}}; t \in \mathcal{T}$$

$$x(i,j,i,l,t) \leq \sum_{p \in \mathcal{PC}(type(i,j))} c^{in}(i,j,i,l,p) y_C(i,j,p) \qquad (5.9)$$

$$\text{for all } (i,j,i,l) \in \mathcal{A}_C \text{ with } (i,l) \in \mathcal{N}_{\mathcal{J}}; t \in \mathcal{T}$$

- Total street capacity constraints:
  The restriction of the total flow for every point in time also depends on the chosen pattern for every street section within an intersection (with $\alpha \geq 2$) and for every street section between two intersections.

$$\sum_{\tilde{t} \in \{t - \tau(i,l,i,h)+1,\dots,t\}} x(i,l,i,h,\tilde{t}) \leq \sum_{p \in \mathcal{PJ}(type(i))} c(i,l,i,h,p) y_{\mathcal{J}}(i,p) \quad (5.10)$$

$$\text{for all } (i,l,i,h) \in \mathcal{A}_{\mathcal{J}}; t \in \mathcal{T}$$

$$\sum_{\tilde{t} \in \{t - \tau(i,l,i,j)+1\dots t\}} x(i,l,i,j,\tilde{t}) \leq \sum_{p \in \mathcal{PC}(type(i,j))} c(i,l,i,j,p) y_C(i,j,p) \quad (5.11)$$

$$\text{for all } (i,l,i,j) \in \mathcal{A}_C \text{ with } (i,l) \in \mathcal{N}_{\mathcal{J}}; t \in \mathcal{T}$$

$$\sum_{\tilde{t} \in \{t - \tau(i,j,i,0)+1\dots t\}} x(i,j,i,l,\tilde{t}) \leq \sum_{p \in \mathcal{PC}(type(i,j))} c(i,j,i,l,p) y_C(i,j,p) \quad (5.12)$$

$$\text{for all } (i,j,i,l) \in \mathcal{A}_C \text{ with } (i,l) \in \mathcal{N}_{\mathcal{J}}; t \in \mathcal{T}$$

Before an instance of the pattern-based evacuation model can be solved, all necessary patterns have to be constructed. In the following, some properties and a construction procedure for patterns within intersections will be presented. The construction procedure for patterns of intersections takes the restrictions presented in Sect. 3.4 into account, but of course, every other restriction may be modeled with the patterns, and, patterns that are not suitable for a certain intersection may be neglected.

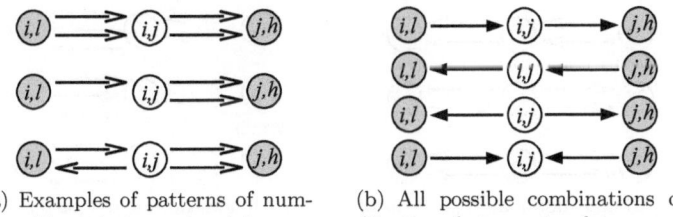

(a) Examples of patterns of number of lanes between i and j

(b) All possible combinations of
directions between i and j

Fig. 5.2 Patterns and combinations of street segments between the two intersections i and j

5.2   Patterns of Street Sections Between Intersections

The construction of all feasible non-dominated patterns of street sections between intersections is less complex compared to the generation of patterns of intersections.

The number of lanes associated with arcs between two intersection has to be determined such that the number of lanes is feasible, i.e. the number of available lanes is not exceeded and every street section is a one-way street. Let i and j be two neighboring intersections, where $(i, j) \in \mathcal{S}$ and $\alpha_i > 1$, $\alpha_j > 1$, then a pattern consists of the number of lanes associated with the arcs (i, l, i, j), (i, j, i, l), (j, h, i, j) and (i, j, j, h), where (i, l) and (j, h) are the appropriate nodes of intersections. A pattern is *non-dominated* if no lane can be added without violating a constraint. Let $\mathcal{PC}$ be the set of all these patterns. Because every street segment is converted into a one-way street, the number of lanes in one of the two opposite directions is zero and the number of lanes in the other one is non- zero, or precisely, it is set to the number of usable lanes in the corresponding direction.

First we will give some examples. In Fig. 5.2a three patterns of a street segment between the intersections i and j are depicted. Let the number of available lanes be two and let an arc represents one lane. The first pattern in Fig. 5.2a (i.e. $\ell(i, l, i, j) = 2$, $\ell(i, j, j, h) = 2$, $\ell(i, j, i, l) = 0$ and $\ell(j, h, i, j) = 0$) is non-dominated in comparison to the shown patterns and it dominates the second one (i.e. $\ell(i, l, i, j) = 1$, $\ell(i, j, j, h) = 2$, $\ell(i, j, i, l) = 0$ and $\ell(j, h, i, j) = 0$). The third pattern is non-dominated, too, but it is not feasible because two-way streets are not permitted.

Following the assumption that all street segments are one-way streets there are just three non-dominated patterns for the street segments that connect two intersections (illustrated in Fig. 5.3a) and there is just one non-dominated pattern for a street segment where one intersection i has just one entrance/exit (i.e. $\alpha_i = 1$).

There are four possible combinations of directions, but just three of them are reasonable. See Fig. 5.2b for an illustration. Here, every arc depicts one direction. The fourth combination of directions in Fig. 5.2b does not lead away from source (i, j), therefore it is not considered.

For every $(i, j) \in \mathcal{P}$, G_C types of patterns are distinguished. These types $type(i, j) \in \{1, \ldots, G_C\}$, $(i, j) \in \mathcal{P}$, denote for example the number of usable

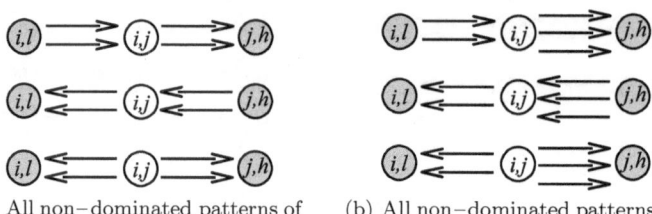

(a) All non–dominated patterns of   (b) All non–dominated patterns of
number of lanes between i and j number of lanes between i and j

Fig. 5.3 Patterns of different types of street segments between the two intersections i and j

lanes or the length of the street segment. The parameters assigned to a pattern are different if the numbers of lanes or the length of a street segment differ. Therefore different types of patterns are introduced. For the determination of patterns (i.e. the numbers of lanes assigned to corresponding arcs) the three feasible and reasonable combinations are crucial. The parameters, like allowed velocity or length of a street section, are given. Figure 5.3 illustrates examples of the three patterns of the connection $(i, j) \in S$ with different types. In Fig. 5.3a on both street segments two lanes are usable. In Fig. 5.3b, there are two lanes usable on the left side and three on the right side. Here, an arc depicts one lane.

Consider now two intersections i and j where $(i, j) \in \mathcal{P}$ and where one intersection has just one entrance/exit, assume $\alpha_i = 1$: If $(i, j) \in S$ then a pattern includes the number of lanes of the arc (i, j, j, h), where (j, h) is the appropriate entrance node to intersection j. If $(i, j) \in \mathcal{D}$ is considered then a pattern includes the number of lanes of the arc (j, h, i, j), where (j, h) is the appropriate exit node of intersection j. In these cases, different types of patterns that distinguish the number of usable lanes are considered as well. But for these last two cases no decision problem is given because just one non-dominated pattern is available.

5.3   Patterns of Intersections

A pattern of an intersection can be translated as the number of lanes of the arcs within the considered intersection. With the number of lanes the parameters inflow capacity and total capacity of arcs can be determined.

The patterns, the number of lanes of an intersection, have to be constructed with respect to some constraints. We decided to choose the same constraints concerning the traffic routing within an intersection as presented in Sect. 3.4 for the one-way urban evacuation model, i.e. the capacity constraint of number of usable lanes, the permission to assign only to one direction of a street segment a positive number of lanes, the prohibition of crossing conflicts and the restriction of merging conflicts. These are the constraints in the one-way urban evacuation model "Lane consistency I" (3.30), "Lane consistency II" (3.32) and the constraints "Prohibition of crossing conflicts I" (3.15) with the coupling constraints (3.28).

The constraints "Prohibition of crossing conflicts II, III" (3.16) and (3.17) are only needed if two-way street are allowed (see Sect. 3.4).

5.3.1 Graphs and Patterns of Intersections

The feasibility of the combination of directions and number of lanes is included in the patterns.

For the ease of notation we will use throughout this section just the number of a node of an intersection instead of two indices for a node, i.e. h instead of (i, h) where h is the number of an entrance/exit and i denotes the considered intersection. That means for the notation of arcs of an intersection, we will use (l, h) instead of (i, l, i, h) to denote an arc of intersection i. A source or a sink will be denoted by a single index as well, i.e. say s instead of (s_1, s_2), where $(s_1, s_2) \in \mathcal{P}$. With that conversion of notation we will repeat existing definitions and add new ones. Directed, undirected and feasible graphs of intersection are defined as follows:

- A (directed) *graph of an intersection* with α entrances/exits is a graph $(\mathcal{N}_\alpha, \mathcal{A}_\alpha)$ with $\mathcal{N}_\alpha = \{1, \ldots, \alpha\}$ and $\mathcal{A}_\alpha = \{(l, h) \in \mathcal{N}_\alpha \times \mathcal{N}_\alpha : l \neq h\}$.
- An *undirected graph of an intersection* is the graph $(\mathcal{N}_\alpha, \mathcal{A}_\alpha)$ where the arcs in $\mathcal{A}_\alpha$ have no orientation (i.e. $(l, h) = (h, l)$). In the undirected case, arcs are also called edges.
- A (directed and undirected) *feasible subgraph* of an intersection with α nodes is a subgraph $(\mathcal{N}'_\alpha, \mathcal{A}'_\alpha)$ of the graph $(\mathcal{N}_\alpha, \mathcal{A}_\alpha)$ of an intersection, where $\mathcal{N}'_\alpha = \mathcal{N}_\alpha$ and $\mathcal{A}'_\alpha \subseteq \mathcal{A}_\alpha$, such that no crossing of arcs/edges occurs.

Examples: Graph of an Intersection

Let us consider an intersection i with four entrances/exits, i.e. $\alpha = 4$. The directed graph of the intersection is depicted in Fig. 5.4a. It contains all twelve possible turn arcs. The undirected graph of i is the complete graph with four nodes, see Fig. 5.4b for an illustration.

A feasible directed graph of intersection i is given in Fig. 5.4c, where the set of arcs is defined as follows: $\mathcal{A}'_\alpha = \{(1, 3), (1, 4), (2, 3)\}$. No pair of these arcs crosses. An infeasible undirected graph of i is shown in Fig. 5.4d with $\mathcal{A}'_\alpha = \{(1, 3), (1, 4), (2, 4)\}$, where the edges $(2, 4)$ and $(1, 3)$ intersect.

Feasibility and domination of patterns of number of lanes of an intersection are defined as follows:

- A *feasible pattern of number of lanes* of an intersection (or a feasible pattern of an intersection) with α nodes consists of the values of number of lanes $\ell(l, h)$, $(l, h) \in \mathcal{A}_\alpha$, where the number of lanes are feasible subject to the constraints of available number of lanes (3.28), the one-way restrictions "Lane consistency I" (3.30), the constraints "Prohibition of crossing conflicts I" (3.15),

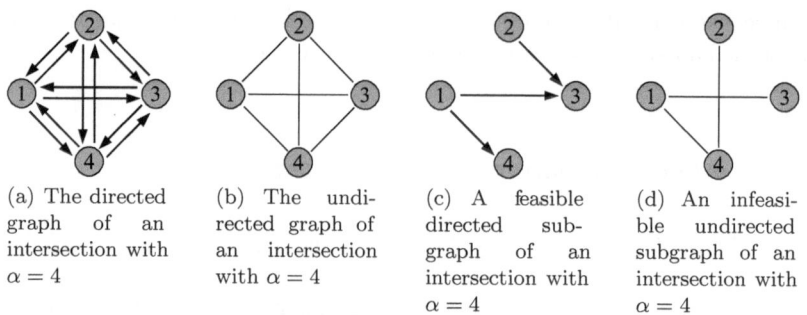

(a) The directed graph of an intersection with $\alpha = 4$

(b) The undirected graph of an intersection with $\alpha = 4$

(c) A feasible directed subgraph of an intersection with $\alpha = 4$

(d) An infeasible undirected subgraph of an intersection with $\alpha = 4$

Fig. 5.4 Examples of graphs of an intersection with four entrances/exits

the constraint of restricting the number of mergings "Lane consistency II" (3.32) and the restriction that every node of the intersection has to be either an entrance or an exit. Otherwise a pattern of number of lanes is called *infeasible*.

- A feasible pattern of number of lanes of an intersection with α nodes is called *non-dominated* if no lane can be added to an arc without violating one of the constraints (3.28), (3.30), (3.15), (3.32) and the restriction that each node of the intersection has to be either an entrance or an exit. It is called *dominated*, if at least one lane can be added without violating one of the considered constraints.
- A pattern p_1 of an intersection is *dominated* by pattern p_2 if for all $(l, h) \in \mathcal{A}_\alpha$ the value $\ell(l, h)$ of p_1 is less than or equal to the value $\ell(l, h)$ of p_2 and at least for one arc $(l, h) \in \mathcal{A}_\alpha$ the inequality is strict.

Based on the pattern of intersections and the parameters of the inflow capacity and total capacity per lane (and per point in time) the parameters inflow and total capacity can be determined for every arc.

Examples: Patterns of Number of Lanes

Let us consider again an intersection i with four entrances/exits, where the maximal number of usable lanes of each direction is set to two. Figure 5.5 illustrates two feasible patterns and two infeasible patterns of number of lanes. Each arc represents an assigned lane.

No crossing occurs in the patterns depicted in Fig. 5.5a, b, the maximal number of lanes is not exceeded and the "merging"-constraint "Lane Consistency II" (3.32) is not violated. The lanes outside the intersections do not belong to the pattern of number of lanes of an intersection. They are needed to determine the feasibility of constraint "Lane Consistency II" (3.32) depending on the number of directions with positive number of lanes and the number of lanes emerging from the considered exit of an intersection. Here both values are equal to two.

Figure 5.5c, d show infeasible patterns of number of lanes. In the pattern shown in Fig. 5.5c a crossing occurs because the directions $(1, 3)$ and $(2, 4)$ both have

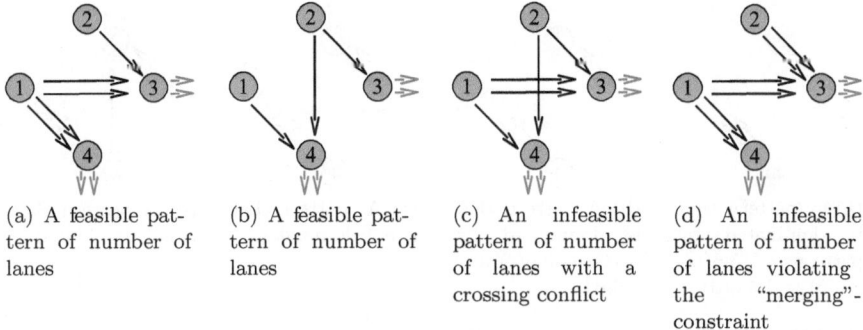

(a) A feasible pattern of number of lanes

(b) A feasible pattern of number of lanes

(c) An infeasible pattern of number of lanes with a crossing conflict

(d) An infeasible pattern of number of lanes violating the "merging"-constraint

Fig. 5.5 Examples of patterns of number of lanes within an intersection with at most two usable lanes in each direction

a positive number of lanes. The pattern of number of lanes depicted in Fig. 5.5d violates the merging-constraint "Lane Consistency II" (3.32) in intersection node 3. The inequality of constraint (3.32) is violated for $\delta = 2$: $\ell(1,3) + \ell(2,3) \not\leq \mathcal{L}(3,s) + \delta - 1$ ($2 + 2 > 2 + 2 - 1$), where $\mathcal{L}(3,s) = 2$ denotes the number of available lanes emerging from exit 3 of the considered intersection to a source or sink s next to exit 3.

Examples: Non-domination and Domination

For the examples of non-dominated and dominated patterns of number of lanes, let us again consider an intersection i with four entrances/exits. Figure 5.6 illustrates examples of non-dominated and dominated patterns of intersection i. Let the maximal usable number of lanes for every street segment within intersection i be two. Patterns 5.6c, d dominate Patterns 5.6a, b. Pattern 5.6a is dominated by all the other depicted Patterns 5.6b–d. Patterns 5.6c, d are non-dominated because no lane can be added without violating the capacity constraint of number of lanes, the non-crossing constraints, the one-way constraints or the constraint restricting the number of mergings.

Note, to the pattern depicted in Fig. 5.6d a lane could be added on arc $(3,2)$ without causing a crossing conflict, exceeding the usable number of lanes or the number of mergings. But if $\ell(3,2) = 1$, then the node 3 of the intersection is an entrance and an exit at the same time. The lane on arc $(3,2)$ would be superfluous because no lane leads from outside the intersection to the intersection node 3.

All feasible and non-dominated patterns $\mathcal{PJ}$ have to be generated before the optimization. Therefore we assume that for a chosen exit l of an intersection the following is valid: $\ell(l,s) = \mathcal{L}(l,s)$ for $(l,s) \in \mathcal{A}_C$ with $s \in \mathcal{S} \cup \mathcal{D}$ and $\mathcal{L}(h,l) \geq \mathcal{L}(l,s)$ for $(h,l) \in \mathcal{A}_J$ and $(l,s) \in \mathcal{A}_C, s \in \mathcal{S} \cup \mathcal{D}$. In a valid pattern, the assigned number of lanes are $\ell(h,l) \leq \mathcal{L}(l,s)$ for $(h,l) \in \mathcal{A}_J$ with $(l,s) \in \mathcal{A}_C$, $s \in \mathcal{S} \cup \mathcal{D}$. (In Chap. 3, we assume that $\mathcal{L}(l_1,l_2) = \mathcal{L}(l_2,l_1)$ for all arcs (l_1,l_2) of

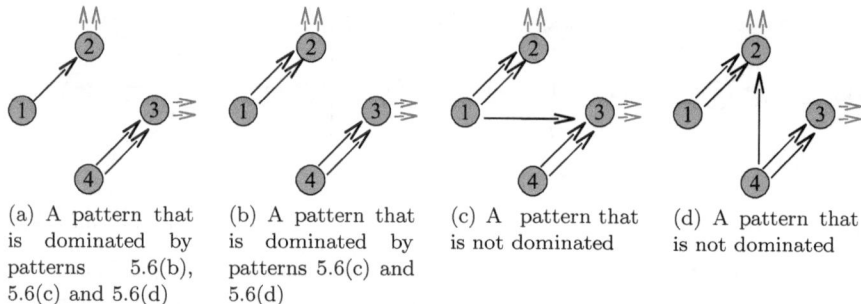

(a) A pattern that is dominated by patterns 5.6(b), 5.6(c) and 5.6(d)

(b) A pattern that is dominated by patterns 5.6(c) and 5.6(d)

(c) A pattern that is not dominated

(d) A pattern that is not dominated

Fig. 5.6 Illustration of dominated and non-dominated patterns within intersection i

the street network. Therefore we have to require that $\mathcal{L}(h, l) \geq \mathcal{L}(l, s)$ for the above mentioned arcs.)

5.3.2   Number of Crossing Conflicts

The number of direction-based crossing conflicts of an intersection with $\alpha \geq 3$ is

$$\alpha (\alpha - 2) + \alpha \sum_{n=0}^{\alpha-3} n (-n + (\alpha - 1)) \qquad (5.13)$$

The crossing conflicts that are considered are modeled "direction-based". With the assumption that every vehicle orders in the appropriate lane that corresponds to the subsequent turn before each vehicle enters an intersection, we just need to consider the crossing of directions. In Fig. 5.7 the crossing conflicts of intersections with $\alpha = 3, 4, 5$ entrances/exits are illustrated. An intersection with $\alpha = 3$ has three crossing conflicts. Each crossing conflict is caused by two left-turn arcs. Intersections with $\alpha = 4$ and $\alpha = 5$ have 16 and 50 possible crossing conflicts, respectively. In Fig. 5.7c, the right turn arcs are not depicted for the sake of clarity.

We count the number of crossing conflicts of an intersection with α entrances/exits by distinguishing three cases: the number of crossing conflicts between two left turn arcs (L/L), the number of crossing conflicts between a left turn arc with a straight forward arc (L/SF) and the number of crossing conflicts of two straight forward arcs (SF/SF). Figure 5.8 illustrates each of the three cases and depicts intersections with $\alpha = 5$ entrances/exits.

L/L   There are α left turn arcs. Consider a fixed left turn arc (l, h) of an intersection, then there are two crossing conflicts with two other left turn arcs that occur, namely $((l - 2) \mod \alpha + 1, l)$ and $(h, h \mod \alpha + 1)$. With α left

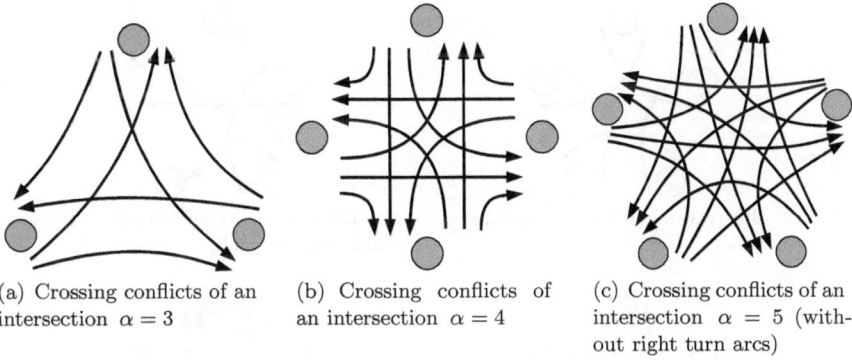

(a) Crossing conflicts of an intersection $\alpha = 3$

(b) Crossing conflicts of an intersection $\alpha = 4$

(c) Crossing conflicts of an intersection $\alpha = 5$ (without right turn arcs)

Fig. 5.7 Illustration of direction-based crossing conflicts of intersections with $\alpha = 3, 4, 5$

turn arcs, there are $\frac{1}{2} \cdot (2\alpha)$ left-left crossing conflicts in total. See Fig. 5.8a for an illustration. All five left-left crossing conflicts are depicted there.

L/SF Consider a fixed left turn arc (l, h) of an intersection. All straight forward arcs (k, g) with $k = h$ or $g = l$ lead to a crossing conflict with (l, h). Thus, all straight forward arcs leaving node h, these are $(\alpha - 3)$ arcs, and all straight forward arcs pointing to node l, these are $(\alpha - 3)$ arcs, cause a crossing conflict. With α left turn arcs, there are $\alpha \cdot (2 \cdot (\alpha - 3))$ crossing conflicts of left turn arcs with straight forward arcs in an intersection with α entrances/exits. See Fig. 5.8b for an illustration.

SF/SF Consider an entrance h of an intersection with α entrances/exits and all $\alpha - 3$ straight forward arcs leaving that entrance. First we will count all crossing conflicts of straight forward arcs that have no node in common: For that intention we consider one crossing straight forward arc after another in counter-clockwise direction. See Fig. 5.8c for an illustration. The first straight forward (h, h_1) arc has one intersection node (node l_1 in Fig. 5.8c) on the right and $(\alpha - 1 - 2)$ on the left (nodes h_2 and l_2 in Fig. 5.8c) that are not head or tail of it. This leads to $2 \cdot (1 \cdot (\alpha - 1 - 2))$ crossing conflicts. The straight forward arc next to the first considered one has two intersection nodes on the right (nodes l_1 and h_1 in Fig. 5.8d) and $(\alpha - 2 - 2)$ on the left (node l_2 in Fig. 5.8d). This leads to $2 \cdot (2 \cdot (\alpha - 2 - 2))$ crossing conflicts. Continue in this manner until the $(\alpha - 3)$th straight forward arc is considered, where $2 \cdot ((\alpha - 3) \cdot (\alpha - (\alpha - 3) - 2)) = 2 \cdot ((\alpha - 3) \cdot 1)$ crossing conflicts occur. In total this leads to $\frac{1}{2}\alpha \sum_{n=1}^{\alpha-3}(2n \cdot (\alpha - (n+2)))$ crossing conflicts of two straight forward arcs that have no node in common.

Crossing conflicts of two straight forward arcs that have a node in common can be counted as follows: Again all straight forward arcs that leave a fixed intersection node are considered one after another counter-clockwise where the first one is the arc on the right. We count the number of intersection nodes on the right of the considered arc. The first arc (arc (h, h_1) in Fig. 5.8e) has one node on the right (node l_1 in Fig. 5.8e), there are zero $(= 2(1 - 1))$ crossing conflicts with a straight forward arc that has a node in common with the considered one because

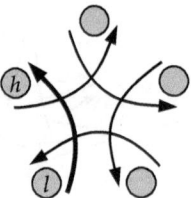

(a) Illustration of left-left crossing conflicts

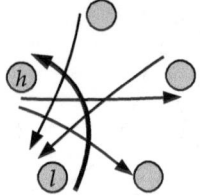

(b) Illustration of left-straight forward crossing conflicts

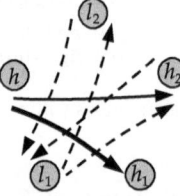

(c) Illustration of crossing conflicts of (h, h_1) with straight forward arcs that have no node in common

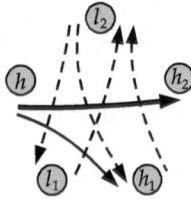

(d) Illustration of crossing conflicts of (h, h_2) with straight forward arcs that have no node in common

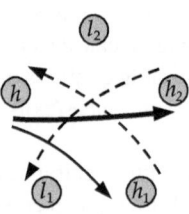

(e) Illustration of crossing conflicts of ctraight forward arcs that have a node in common

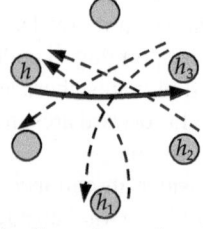

(f) Illustration of crossing conflicts of straight forward arcs that have no node in common of an intersection with $\alpha = 6$

Fig. 5.8 Illustration of crossing conflicts of intersections with $\alpha = 5$ and $\alpha = 6$

there no straight forward arc exists that points or emerges from the node on the right to the first considered arc. These arcs are "just" left and right turn arcs (in Fig. 5.8e these are the left turn arcs (h_1, l_1), (l_1, h) and the right turn arcs (h, l_1) and (l_1, h_1)). The second straight forward arc with two intersection nodes on the right crosses with two $(= 2(2 - 1))$ straight forward arcs that, each, have a node in common with the considered one (the bold straight forward arc (h, h_2) in Fig. 5.8e with (h_1, h) and (h_2, l_1)). Figure 5.8f illustrates an intersection with $\alpha = 6$ entrances/exits. Entrance h is considered and the third straight forward arc (h, h_3) counted from the most right one is marked in bold. With arc (h, h_3) four $(= 3(3 - 1))$ straight forward arcs cross that share a node with it.

Table 5.1 Number of crossing conflicts of intersections with $\alpha = 3, \ldots, 7$ entrances/exits

α	3	4	5	6	7
No. of CC's	3	16	50	120	245

That means the nth considered straight forward arc, i.e. the nth straight forward arc counted from the most right one counter-clockwise, with n intersection nodes on the right leads to $2 \cdot (n-1)$ crossing conflicts of the considered kind. Overall with α entrances/exits and $\alpha - 3$ straight forward arcs emerging from an intersection node, there are $\frac{1}{2}\alpha \sum_{n=1}^{\alpha-3} 2(n-1)$ crossing conflicts of two straight forward arcs that have a node in common within an intersection with α entrances/exits.

All numbers of crossing conflicts summed up lead to the result (5.13) from above:

$$\alpha + 2\alpha(\alpha - 3) + \frac{1}{2}\alpha \sum_{n=1}^{\alpha-3}(2n \cdot (\alpha - (n+2)) + 2(n-1))$$

$$= \alpha + 2\alpha(\alpha - 3) + \alpha \sum_{n=1}^{\alpha-3}(n\alpha - n^2 - n - 1)$$

$$= (\alpha + 2\alpha^2 - 6\alpha) - \alpha(\alpha - 3) + \alpha \sum_{n=1}^{\alpha-3}(n\alpha - n^2 - n)$$

$$= \alpha + 2\alpha^2 - 6\alpha - \alpha^2 + 3\alpha + \alpha \sum_{n=1}^{\alpha-3}(n\alpha - n^2 - n)$$

$$= \alpha^2 - 2\alpha + \alpha \sum_{n=1}^{\alpha-3} n(-n + (\alpha - 1))$$

In Table 5.1 the number of direction-based crossing conflicts of intersections with $\alpha = 3, \ldots, 7$ entrances/exits are listed.

To illustrate the crossing conflicts of an intersection with $\alpha = 5$ entrances and exits in Fig. 5.9, each crossing conflict is marked with a node. Right turn arcs are not illustrated in Fig. 5.9 for the sake of clarity. They cause no crossing conflict. The dark gray nodes are crossing conflicts between two left turn arcs, the light gray nodes represent crossing conflicts of a left turn arc with a straight forward arc and the white nodes represent crossing conflicts with two straight forward arcs. First the dark gray nodes are counted and numbered, then the light gray nodes and finally the white nodes are counted and numbered clockwisely.

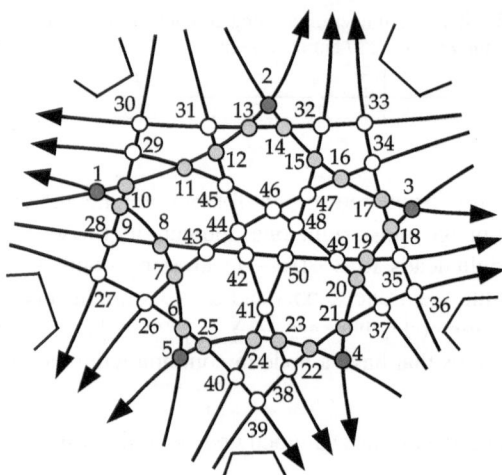

Fig. 5.9 Illustration of direction-based crossing conflicts of an intersection with $\alpha = 5$

5.3.3 Planar Graphs and Patterns

With the help of planar graphs, an upper bound on the number of feasible subgraphs of an intersection can be determined. Some properties of planar graphs can be used for solving the problem of a crossing-conflict-free undirected subgraph of an intersection.

A *planar* graph is a graph (see e.g. Clark and Holton 1995, p. 157) that can be drawn on the plane without intersecting edges.

We will use the following lemma concerning the number of edges of a planar graph (see e.g. Clark and Holton, 1995, p.165): Let n be the number of nodes, then the number of edges e of a planar graph with $n \geq 3$ is bounded by

$$e \leq 3n - 6 \qquad (5.14)$$

The goal is to construct feasible subgraphs of intersections based on planar graphs.

Example: We will start with an example. Consider a planar graph with five nodes $\{1, \ldots, 5\}$ and the set of edges $\{(1, 3), (1, 4), (1, 5), (2, 3)\}$, illustrated in Fig. 5.10a. The directions of the turns have to be included to the planar graph to build a feasible directed subgraph of an intersection with $\alpha = 5$. The first example is given in Fig. 5.10b. Let the following directions of the edges be given: $(3, 1), (4, 1), (5, 1)$ and $(3, 2)$. With these turn directions a feasible subgraph (see Fig. 5.10c for an illustration) can be generated based on the planar graph considered.

But note, the resulting subgraph of an intersection based on an planar graph will not always be crossing-free if we considered two-way streets. If the directions of the considered planar graph are $(1, 3), (4, 1), (5, 1)$ and $(2, 3)$ (see Fig. 5.11b), then

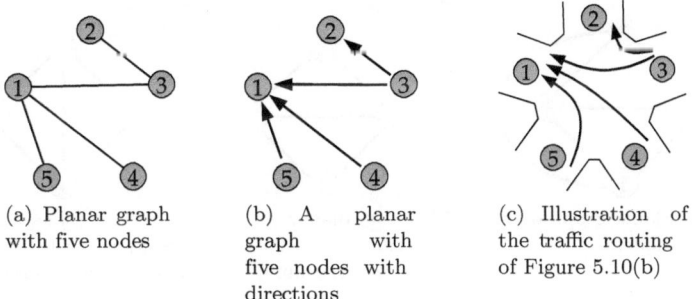

(a) Planar graph with five nodes

(b) A planar graph with five nodes with directions

(c) Illustration of the traffic routing of Figure 5.10(b)

Fig. 5.10 Illustration of a feasible traffic routing within an intersection based on a planar graph

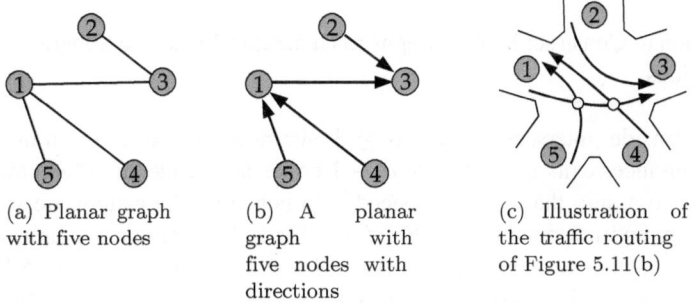

(a) Planar graph with five nodes

(b) A planar graph with five nodes with directions

(c) Illustration of the traffic routing of Figure 5.11(b)

Fig. 5.11 Illustration of an infeasible traffic routing within an intersection based on a planar graph

there are two crossing conflicts in the resulting directed subgraph (see Fig. 5.11c): $(1, 3)$ crosses $(5, 1)$ and $(4, 1)$.

With the consideration of the two-way streets, we have to be aware that not all possible crossing conflicts are detected with the approach of the planar graphs. All left-left and left-straight forward crossing conflicts as well as a crossing conflict of two straight forward arcs that share a node remain undetected. In the two-way case, an access to an intersection can be an entrance and an exit of the intersection at the same time. That causes the above mentioned crossing conflicts to occur. If just one-way streets are allowed, then just crossing conflicts of straight forward arcs that have no node in common may occur.

In our instances we restrict to one-way streets and hence on patterns that represent one-way streets. If every street section is readjusted to a one-way street, then every node of an intersection is either an entrance or an exit. As a consequence the directions of the edges of the subgraph of an intersection based on a planar graph are determined. Therefore, we disregard the consideration of directions in the next steps.

Another conclusion learned from the observations of the examples illustrated in Figs. 5.10 and 5.11 is that a planar undirected subgraph of an intersection can lead to different traffic routings depending on which nodes of the intersection are chosen as entrances and which as exits.

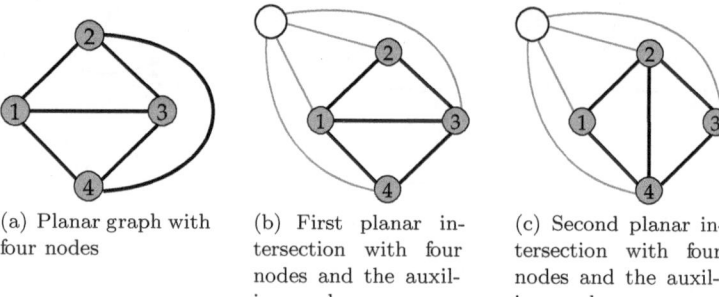

(a) Planar graph with four nodes

(b) First planar intersection with four nodes and the auxiliary node

(c) Second planar intersection with four nodes and the auxiliary node

Fig. 5.12 Illustration of building a planar graph for a crossing-free intersection

Preparation to Construct Planar Graphs to Generate all Feasible Subgraphs of Intersections

We will consider intersections with $\alpha \geq 3$ entrances and exits. For an intersection with α entrances/exits a graph with $\alpha + 1$ nodes is considered. The extra node is necessary to assure that the edges needed to construct the pattern are within the intersection and not outside. See Fig. 5.12a for an illustration: a graph with four nodes has equal or less than six edges ($3 \cdot 4 - 6 = 6$, see inequality (5.14)). The edge $(2, 4)$ would be outside the intersection and would cross the street section that connects the considered intersections with a neighboring source or sink, if we interpreted the graph as an intersection with four entrances/exits.

For that reason an auxiliary node is added to the α intersection nodes. All α nodes of the intersection are connected to the auxiliary node by edges. Hence, the maximal number of edges of an intersection with α nodes and the auxiliary node that build a planar graph is the following (using inequality (5.14)):

$$e \leq 3(\alpha + 1) - 6 = 3(\alpha - 1) \qquad (5.15)$$

Every node of the intersection is connected to the auxiliary node, hence the number $e_{\mathcal{J}}$ of edges that remain to be allocated to a subgraph of the intersection graph is the following:

$$e_{\mathcal{J}} \leq 3(\alpha - 1) - \alpha = 2\alpha - 3 \qquad (5.16)$$

Figure 5.12b, c illustrate the planar graphs that can be generated for an intersection with $\alpha = 4$. The number of edges of an planar graph with $\alpha + 1$ nodes is at most $3(\alpha - 1)$, in this case nine. Four of these edges are connected to the auxiliary node (the gray edges in Fig. 5.12b, c). Hence five ($2\alpha - 3 = 5$) edges remain to build a planar graph. There are two possibilities to complete the planar graph, shown in Fig. 5.12b, c. Hence we have two possible subgraphs of an intersection with four entrances/exits where the directions are not yet considered.

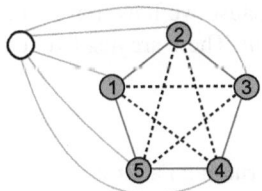

Fig. 5.13 Illustration of building a planar graph for a crossing-free intersection

To determine an upper bound for the number of such graphs the following further considerations can be made: The edges $(l, (l \mod \alpha) + 1)$, $l \in \{1, \ldots, \alpha\}$ at the boundary of the intersection, i.e. the edges that represent a left or a right turn arcs later, can be chosen directly because they do not lead to crossing conflicts in the one-way case. Then the number $e_{\mathcal{J}}^{inner}$ of "inner" edges of the intersection that remain to be chosen are

$$e_{\mathcal{J}}^{inner} \le 2\alpha - 3 - \alpha = \alpha - 3, \tag{5.17}$$

i.e. the edges that represent straight forward arcs later. Right turn arcs never cause a crossing conflict, left arcs just cause crossing conflicts if two-way traffic is allowed. Remember, a left turn arc is understood as an arc that points directly to the neighboring node to the left of the node it is emerging from. We refer to the realization of the corresponding constraints that prohibit the crossing conflicts of left turn and straight forward arcs that build a crossing conflict and that share a node (see constraints (3.16) and (3.17)).

Let us consider the example of an intersection with $\alpha = 5$ entrances/exits, see Fig. 5.13 for an illustration. There are $3(\alpha - 1) = 12$ edges to construct a planar graph with six nodes. Five edges are allocated from the node of the intersection to the auxiliary node (the light gray arcs) and the five edges $(1, 2)$, $(2, 3)$, $(3, 4)$, $(4, 5)$ and $(5, 1)$ within the intersection are allocated (dark gray arcs). Then there are at most two arcs that can be chosen from the edges in the inner of the intersection, the dashed black arcs ($e_{\mathcal{J}}^{inner} \le 2$, see inequality (5.17)). Hence there are at most $\binom{5}{2} = 10$ pattern of feasible edges.

A general upper bound of the number of feasible undirected subgraphs of an intersection with $\alpha \ge 3$ entrances/exits is

$$\binom{\frac{1}{2}(\alpha(\alpha - 3))}{\alpha - 3} \tag{5.18}$$

where $\frac{1}{2}(\alpha(\alpha - 3))$ is the number of all "inner" edges of an intersection with α entrances/exits. Note, the entrances and exits are not yet chosen and different choices of entrances and exits of an intersection lead to different feasible directed subgraphs of an intersection.

The bound is not tight because we have not included that crossing conflicts of straight forward arcs may occur. There are just five feasible undirected subgraphs of an intersection with $\alpha = 5$.

A General Approach of Constructing Patterns

First, all possible combinations of entrances/exits of an intersection are determined to construct all feasible patterns of an intersection. They are considered separately. Based on a choice of entrances and exits, feasible and reasonable edges are determined, i.e. all feasible subgraphs of the intersection with the considered entrance/exit pattern. Based on the feasible subgraphs and the available number of lanes, all patterns of number of lanes of the considered intersection can be determined.

The general approach to construct patterns is the following

1. Build all patterns of entrances/exits.
2. Generate all feasible directed subgraphs (which arcs of the intersection graph $(\mathcal{N}_\alpha, \mathcal{A}_\alpha)$ can have an assignment of a positive number of lanes at the same time) for all patterns of entrances/exits. The constraints "Prohibition of crossing conflicts I" (3.15) and one-way constraint (3.30) are taken into account.
3. Generate all feasible patterns of number of lane based on the feasible directed subgraphs of the intersection. Constraints concerning the number of lanes, namely capacity of number of lanes (3.28) and the merging- constraint "Lane consistency II" (3.32), are taken into account.

Patterns of Entrances/Exits

A pattern of entrances/exits of an intersection with α entrances/exits is the determination which nodes of the intersection are entrances and which nodes are exits. The generation of patterns of entrances/exits is nothing but the determination of all subsets of the set $\mathcal{N}_\alpha = \{1, \ldots, \alpha\}$ without the empty set and the set $\mathcal{N}_\alpha$ itself. Therefore the total number of entrance/exit patterns is

$$\sum_{i=1}^{\alpha-1} \binom{\alpha}{i}$$

(choosing i entrances (w.l.o.g. otherwise exits) out of α nodes, $i = 1, \ldots, (\alpha - 1)$). They can be generated with binary numbers with α digits.

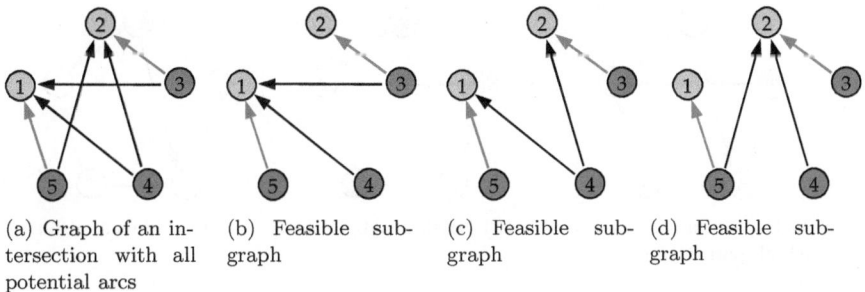

(a) Graph of an intersection with all potential arcs   (b) Feasible subgraph   (c) Feasible subgraph   (d) Feasible subgraph

Fig. 5.14 Illustration of directed subgraphs of an intersection

Subgraphs of Intersections

The directions within an intersection that can have a positive number of lanes simultaneously such that it does not violate the crossing conflict constraints can be determined with planar graphs. The arcs with a positive number of lanes are the arcs of a feasible subgraph of an intersection.

Let a pattern of entrances/exits of an intersection with α nodes be given, where $\mathcal{N}_{entr} \subset \mathcal{N}_{\alpha}$ denotes the set of entrances and $\mathcal{N}_{exit} \subset \mathcal{N}_{\alpha}$ denotes the set of exits of the considered entrance/exit pattern with $\mathcal{N}_{entr} \cup \mathcal{N}_{exit} = \mathcal{N}_{\alpha}$ and $\mathcal{N}_{entr} \cap \mathcal{N}_{exit} = \emptyset$. The number of straight forward arcs of a feasible subgraph of an intersection with α nodes does not exceed $\alpha - 3$ (inequality (5.17)). The set of feasible arcs of an entrance/exit pattern are all arcs that lead from an entrance to an exit of the intersection, i.e. $\mathcal{N}_{entr} \times \mathcal{N}_{exit}$. The arcs in $\mathcal{N}_{entr} \times \mathcal{N}_{exit}$ build all potential arcs of a feasible subgraph of an intersection.

The computation of the upper bound of number of different undirected subgraphs given above (see (5.18)) cannot be adopted for every specific entrance/exit pattern, i.e. if directions are taken into account. It does not stay valid if strictly less than $\alpha - 3$ arcs can be chosen to generate a feasible directed subgraph. For an illustration, two examples are given in the following.

Example: $\alpha - 3$ Straight Forward Arcs can be Chosen

Let us consider an intersection with $\alpha = 5$ and the entrance/exit pattern $\mathcal{N}_{entr} = \{3, 4, 5\}, \mathcal{N}_{exit} = \{1, 2\}$. In Fig. 5.14 the dark gray nodes represent entrances and the light gray nodes exits. The set of potential arcs of the subgraph is $\{3, 4, 5\} \times \{1, 2\}$. There are at most $\alpha - 3$ (inequality (5.17)) straight forward arcs that can be chosen to achieve a feasible subgraph. The gray arrows depict the left and right turn arcs in Fig. 5.14a. They do not cause a crossing conflict and are therefore within the set of arcs of every feasible subgraph. The black arrows depict the straight forward arcs, from which at most $2 (= \alpha - 3)$ arcs have to be selected. There are four straight forward arcs, hence $\binom{4}{2} = 6$ possibilities to choose two arcs. This is an upper bound

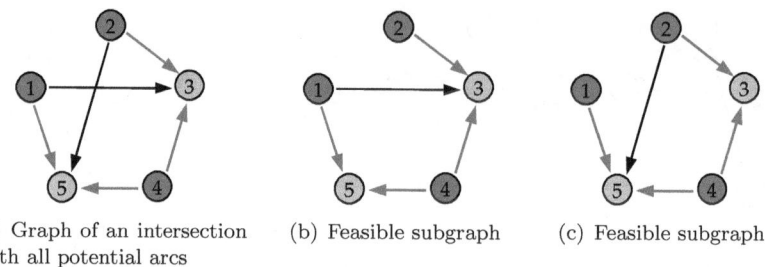

(a) Graph of an intersection   (b) Feasible subgraph   (c) Feasible subgraph
with all potential arcs

Fig. 5.15 Illustration of directed subgraphs of an intersection

of the number of feasible subgraphs that are based on the given entrance/exit pattern. The number of feasible subgraphs is three, depicted in Fig. 5.14b–d.

Example: Only Less than $\alpha - 3$ Straight Forward Arcs can be Chosen

The second example is chosen as follows: Consider an intersection with $\alpha = 5$ entrances/exits and let the entrance/exit pattern be $\mathcal{N}_{entr} = \{1, 2, 4\}$, $\mathcal{N}_{exit} = \{3, 5\}$. The example is illustrated in Fig. 5.15. There are six arcs that can be chosen to construct a feasible subgraph: $(1, 3)$, $(1, 5)$, $(2, 3)$, $(2, 5)$, $(4, 3)$ and $(4, 5)$. Left and right turn arcs do not lead to crossing conflicts if every node of an intersection is either an entrance or an exit, like in the considered case. Hence the arcs $(1, 5)$, $(2, 3)$, $(4, 3)$ and $(4, 5)$ (represented as gray arrows in Fig. 5.15a) are in every subgraph without causing any crossing conflict. With the same arguments as before, there are, at most, two straight forward arcs that can be chosen to achieve a feasible subgraph. In this example there are just two straight forward arcs left to choose from. But they intersect, i.e. they cannot be chosen both in one pattern. The consideration that $\binom{2}{2} = 1$ is an upper bound for the number of feasible subgraphs is invalid. There are two feasible subgraphs of the intersection with the given pattern of entrances/exits. They are illustrated in Fig. 5.15b, c. The upper bound of the number of feasible undirected subgraphs of an intersection with $\alpha \geq 3$ entrances/exits does not stay valid if directions in terms of entrance and exit nodes are included and therefore only a subset of edges/arcs of all possible straight forward arcs can be chosen.

Construction of Feasible Directed Subgraphs

A method to construct all feasible directed subgraphs based on a pattern of entrances/exits of an intersection will be presented in the following. Therefore following sets are defined:

- Let $CC(l, h)$ be the set of straight forward arcs that cross with $(l, h) \in (\mathcal{N}_{entr} \times \mathcal{N}_{exit}) \cap \mathcal{A}_{\alpha}^{sf}$. $CC(l, h)$ is determined by the following set:

$$\left((\mathcal{N}_{entr} \cap SF^{\alpha}_{(l,h)}) \times (\mathcal{N}_{exit} \cap SF^{\alpha}_{(h,l)}) \right) \cup \left((\mathcal{N}_{entr} \cap SF^{\alpha}_{(h,l)}) \times (\mathcal{N}_{exit} \cap SF^{\alpha}_{(l,h)}) \right)$$

where $SF^{\alpha}_{(l,h)} = \{k_a, \ldots, k_e\}$ if $k_a \le k_e$ and $SF^{\alpha}_{(l,h)} = \{1, \ldots, \alpha\} \backslash \{k_e + 1, \ldots, k_a - 1\}$ if $k_a > k_e$ with $k_a = (l \mod \alpha) + 1$, $k_e = \alpha - ((\alpha - h + 1) \mod \alpha)$ (set SF is already defined for constraint 3.15).

- Let $CC := \{(l,h) \in \mathcal{N}_{entr} \times \mathcal{N}_{exit} : |CC(l,h)| > 0\}$ be the set of arcs that cause at least one crossing conflict.
- Let $\overline{CC}(l,h) := CC \backslash (CC(l,h) \cup \{(l,h)\})$ for all $(l,h) \in CC$ be the set of arcs that do not cross with (l,h).
- Let $CH(P(l,h))$ be the set of children of the parent node $P(l,h)$ of node (l,h).

The algorithm generates trees composed of straight forward arcs. A tree is organized in levels: starting in level 1 a root node and ending in level $\alpha - 3$. The nodes of the trees consist of straight forward arcs that do not cross with the root node. There are at most $\alpha - 3$ straight forward arcs to be chosen to construct a crossing-free subgraph including the root node (it is associated with an arc) in level 1. The elements of paths from a root to a leave include straight forward arcs that do not build a crossing conflict. With Pseudocode 7 all feasible directed subgraphs are generated, possibly multiple times. Two examples will be given below. The algorithm is reasonable to consider for intersections with more than four entrances/exits, i.e. $\alpha \ge 5$. In intersections with $\alpha = 3$ no straight forward arc exists, hence crossing conflicts of two straight forward arcs cannot occur. For intersections with $\alpha = 4$, just one straight forward arc needs to be chosen.

Pseudocode 7 Generation of feasible subsets of straight forward arcs for intersections with $\alpha \ge 5$

```
 1: for (l_0, h_0) ∈ CC do
 2:     Node of level 1: root node (l_1, h_1)
 3:     Nodes of level 2: children (l_2, h_2) of root node (l_1, h_1) are all arcs in CC̄(l_1, h_1)
 4:     if α ≥ 6 then
 5:         for k = 3, ... (α − 3) do
 6:             for (l_{k−1}, h_{k−1}) is a node of level k − 1 and ≠ END do
 7:                 if CC̄(l_{k−1}, h_{k−1}) ∩ CH(P(l_{k−1}, h_{k−1})) ≠ ∅ then
 8:                     Nodes of level k: children (l_k, h_k) of arc (l_{k−1}, h_{k−1}) are all arcs in
                        CC̄(l_{k−1}, h_{k−1}) ∩ CH(P(l_{k−1}, h_{k−1}))
 9:                 else
10:                     Child of (l_{k−1}, h_{k−1}) is called END
11:                 end if
12:             end for
13:         end for
14:     end if
15: end for
```

The elements of level k are indicated by k. For every arc $(l,h) \in CC$ a separate tree is generated. The directed subgraphs are built by the arcs (l,h) within $\mathcal{N}_{entr} \times \mathcal{N}_{exit}$ with $CC(l,h) = \emptyset$ and the arcs being components of a path from

Table 5.2 Sets CC and $\overline{CC}$ of the arcs of an intersection with $\alpha = 5$ entrances/exits

Arc (l, h)	$CC(l, h)$	$\overline{CC}(l, h)$
$(3, 1)$	$(4, 2), (5, 2)$	$(4, 1)$
$(3, 2)$	$\emptyset$	–
$(4, 1)$	$(5, 2)$	$(3, 1), (4, 2)$
$(4, 2)$	$(3, 1)$	$(4, 1), (5, 2)$
$(5, 1)$	$\emptyset$	–
$(5, 2)$	$(3, 1), (4, 1)$	$(4, 2)$

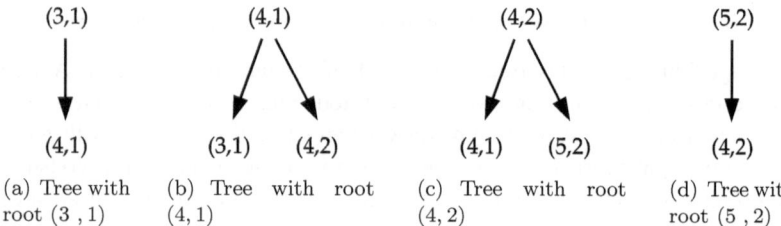

(a) Tree with root $(3, 1)$ (b) Tree with root $(4, 1)$ (c) Tree with root $(4, 2)$ (d) Tree with root $(5, 2)$

Fig. 5.16 Illustration of the trees generated with Pseudocode 7

a root $(l_1, h_1) \in CC$ to a leaf of the corresponding tree. If $\overline{CC}(l_{k-1}, h_{k-1}) \cap CH(P(l_k, h_k)) = \emptyset$ then the considered path ends with node (l_{k-1}, h_{k-1}) consisting of a straight forward arc and the child of (l_{k-1}, h_{k-1}), the node END, to map the end of the path. The set $CH(P(l_2, h_2))$ is equal to the set of arcs that do not cross with the corresponding root node (l_1, h_1) $(CH(P(l_2, h_2)) = CH(l_1, h_1) = \overline{CC}(l_1, h_1)$ per construction, see Pseudocode 7 Line 3). Next, the algorithm will be illustrated with two examples.

Example: An Intersection with $\alpha = 5$ Entrances/Exits

The pattern of entrances/exits is defined by $\mathcal{N}_{entr} = \{3, 4, 5\}$, $\mathcal{N}_{exit} = \{1, 2\}$. The left and right turn arcs $(3, 2)$ and $(5, 1)$ with no crossing conflicts belong to every feasible subgraph. The directed graph with all feasible arcs corresponding to the entrance/exit pattern is depicted in Fig. 5.14a. The sets that have to be considered for the algorithm are depicted in Table 5.2. The set of arcs that cause at least one crossing conflict is $CC = \{(3, 1), (4, 1), (4, 2), (5, 2)\}$. See Fig. 5.16 for an illustration of the trees of the different roots $(l, h) \in CC$. Just Lines 1–3 of Pseudocode 7 need to be executed because $\alpha = 5$ and hence just two straight forward arcs have to be chosen.

The resulting different paths of straight forward arcs are $(3, 1)$–$(4, 1)$, $(4, 1)$–$(4, 2)$ and $(4, 2)$–$(5, 2)$. The results lead to three different feasible directed subgraphs with the following sets of arcs: $\mathcal{A}_5^1 = \{(3, 2), (5, 1)\} \cup \{(3, 1), (4, 1)\}$, $\mathcal{A}_5^2 = \{(3, 2), (5, 1)\} \cup \{(4, 1), (4, 2)\}$ and $\mathcal{A}_5^3 = \{(3, 2), (5, 1)\} \cup \{(4, 2), (5, 2)\}$. The feasible directed subgraphs of the considered intersection are depicted in Fig. 5.14b–d.

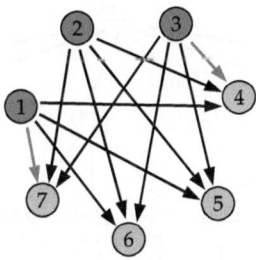

Fig. 5.17 An illustration of a graph of an intersection ($\alpha = 7$) with all potential arcs

Table 5.3 Sets CC and $\overline{CC}$ of the arcs of an intersection with $\alpha = 7$ nodes

Arc (l, h)	$CC(l, h)$	$\overline{CC}(l, h)$
$(1, 4)$	$(2, 5), (2, 6), (2, 7), (3, 5),$ $(3, 6), (3, 7)$	$(1, 5), (1, 6), (2, 4)$
$(1, 5)$	$(2, 6), (2, 7), (3, 6), (3, 7)$	$(1, 4), (1, 6), (2, 4), (2, 5), (3, 5)$
$(1, 6)$	$(2, 7), (3, 7)$	$(1, 4), (1, 5), (2, 4), (2, 5),$ $(2, 6), (3, 5), (3, 6)$
$(1, 7)$	$\emptyset$	$-$
$(2, 4)$	$(3, 5), (3, 6), (3, 7)$	$(1, 4), (1, 5), (1, 6), (2, 5),$ $(2, 6), (2, 7)$
$(2, 5)$	$(1, 4), (3, 6), (3, 7)$	$(1, 5), (1, 6), (2, 4), (2, 6),$ $(2, 7), (3, 5)$
$(2, 6)$	$(1, 4), (1, 5), (3, 7)$	$(1, 6), (2, 4), (2, 5), (2, 7),$ $(3, 5), (3, 6)$
$(2, 7)$	$(1, 4), (1, 5), (1, 6)$	$(2, 4), (2, 5), (2, 6), (3, 5),$ $(3, 6), (3, 7)$
$(3, 4)$	$\emptyset$	$-$
$(3, 5)$	$(1, 4), (2, 4)$	$(1, 5), (1, 6), (2, 5), (2, 6),$ $(2, 7), (3, 6), (3, 7)$
$(3, 6)$	$(1, 4), (1, 5), (2, 4), (2, 5)$	$(1, 6), (2, 6), (2, 7), (3, 5), (3, 7)$
$(3, 7)$	$(1, 4), (1, 5), (1, 6), (2, 4),$ $(2, 5), (2, 6)$	$(5, 2), (6, 2), (6, 4), (7, 4)$

Example: An Intersection with $\alpha = 7$ Entrances/Exits

The algorithm is more interesting for intersections with a higher number of nodes. Here the drawback of the algorithm becomes apparent; several feasible subgraphs are generated multiple times.

Let us consider an intersection with $\alpha = 7$ entrances/exits and the pattern of entrances/exits defined by $\mathcal{N}_{entr} = \{1, 2, 3\}$ and $\mathcal{N}_{exit} = \{4, 5, 6, 7\}$. An illustration of the graph of the intersection with all potential arcs is given Fig. 5.17.

The right turn arc $(1, 7)$ and the left turn arc $(3, 4)$ are part of every feasible subgraph. The sets $CC(l, h), \overline{CC}(l, h)$ are shown in Table 5.3. The set CC of arcs that cross at least with another straight forward arc consists of the following arcs:

$$CC = \{(1, 4), (1, 5), (1, 6), (2, 4), (2, 5), (2, 6), (2, 7), (3, 5), (3, 6), (3, 7)\}.$$

Hence, ten trees have to be built with Pseudocode 7.

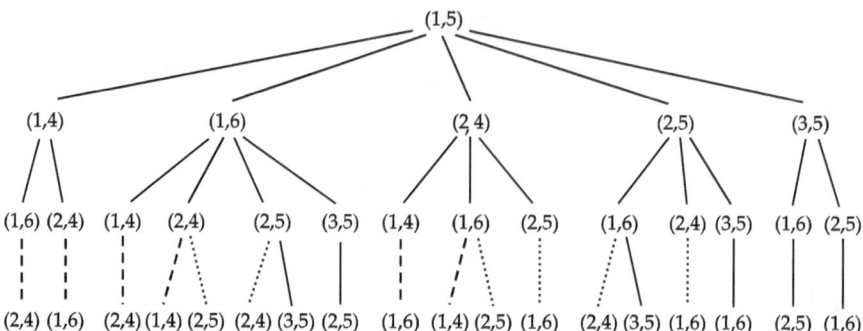

Fig. 5.18 An illustration of the tree determined with Pseudocode 7 with root $(1, 5)$

We will consider the for-loop for the element $(1, 5)$ of the set CC. For an illustration of the resulting tree with root $(1, 5)$ see Fig. 5.18. The first level is the root node, it consists of the chosen arc $(1, 5)$. The second level consists of the elements of the set $\overline{CC}(1, 5) = \{(1, 4), (1, 6), (2, 4), (2, 5), (3, 5)\}$. Each of the elements is a child of root $(1, 5)$ of the considered tree of straight forward arcs. Then the third level has to be determined. I.e. in Pseudocode 7 the for-loop in Line 5 is executed for $k = 3$. The children of the first node $(1, 4)$ of the second level are the arcs in set $\overline{CC}(1, 4) \cap CH(P(1, 4)) = \overline{CC}(1, 4) \cap \overline{CC}(1, 5) = \{(1, 6), (2, 4)\}$, the children of the second node $(1, 6)$ of the second level are the arcs in set $\overline{CC}(1, 6) \cap \overline{CC}(1, 5) = \{(1, 4), (2, 4), (2, 5), (3, 5)\}$. The children of nodes $(2, 4)$, $(2, 5)$ and $(3, 5)$ are determined analogously. The fourth level has to be determined: The for-loop (Line 5 in Procedure 7) for $k = 4 (= \alpha - 3)$ is executed. The children of the first node $(1, 6)$ of the third level are the arcs in set $\overline{CC}(1, 6) \cap CH(P(1, 6)) = \overline{CC}(1, 6) \cap \overline{CC}(1, 4) \cap \overline{CC}(1, 5) = \{(2, 4)\}$, the children of the second node $(2, 4)$ of the third level are the arcs in set $\overline{CC}(2, 4) \cap \overline{CC}(1, 4) \cap \overline{CC}(1, 5) = \{(1, 6)\}$. The children for the third node $(1, 4)$ of the third level are the arcs in set $\overline{CC}(1, 4) \cap \overline{CC}(1, 6) \cap \overline{CC}(1, 5) = \{(2, 4)\}$ and for the fourth node $(2, 4)$ of the third level the children are the arcs in set $\overline{CC}(2, 4) \cap \overline{CC}(1, 6) \cap \overline{CC}(1, 5) = \{(1, 4), (2, 5)\}$. For the nodes $(2, 5)$, $(3, 5)$, $(1, 4)$, $(1, 6)$, $(2, 5)$, ..., $(2, 5)$ of the third level the child nodes are determined in the same manner. The procedure ends with the computation of the fourth level.

Three different pattern are determined with the tree of root node $(1, 5)$. These are the patterns that consists of the set of arcs

$$\{(1, 4), (1, 5), (1, 6), (2, 4)\} \cup \{(1, 7), (3, 4)\},$$

$$\{(1, 5), (1, 6), (2, 4), (2, 5)\} \cup \{(1, 7), (3, 4)\} \text{ and}$$

$$\{(1, 5), (1, 6), (2, 5), (3, 5)\} \cup \{(1, 7), (3, 4)\}$$

In Fig. 5.18, the different sets of straight forward arcs that do not cross are highlighted with dashed, dotted and solid lines between the third and fourth level. In total, there are ten patterns.

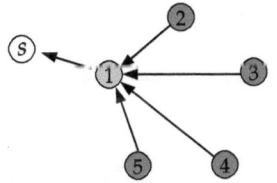

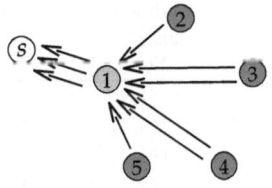

(a) Feasible subgraph of an intersection with four entrances and one exit

(b) Feasible pattern of number of lanes of an intersection with four entrances and one exit

Fig. 5.19 Illustration of a feasible subgraph and pattern of number of lanes of an intersection with $\alpha = 5$

Patterns of Number of Lanes

If a feasible subgraph of an intersection is given, all patterns of number of lanes based on the given subgraph can be determined by means of the "combinations of integer numbers".

The number of lanes needs to be determined such that the generated pattern of number of lanes is feasible, i.e. it has to be assured that the number of usable lanes is not exceeded (modeled with constraint (3.28)) and that the "merging"-constraint "Lane consistency II" (3.32) is not violated. To generate all feasible numbers of lanes the k-composition of integer numbers can be used (see e.g. Stanley, 1997, p. 14): A k-composition of an integer n with k integer numbers ($n \geq k$) is: $x_1 + x_2 + \ldots + x_k = n$ where $x_i \geq 1, i = 1, \ldots, k$. There are $\binom{n-1}{k-1}$ k-compositions of n with k parts. The idea of the proof of the number of k-compositions is the following: there are n balls that need to be separated by $k - 1$ lines. The separation lines can be placed on $n - 1$ places, hence there are $\binom{n-1}{k-1}$ possibilities to solve the equation.

Example: Let us consider an intersection with $\alpha = 5$ and four entrance nodes 2, 3, 4 and 5 (represented as dark gray nodes in Fig. 5.19) and one exit node 1 (depicted as light gray node in Fig. 5.19). Figure 5.19 illustrates this example. The filled arrows depict the arcs of a directed subgraph (Fig. 5.19a), each non-filled arrow depicts one lane (Fig. 5.19b).

Let the set of arcs of the subgraph be $\mathcal{A}_\alpha = \{(2, 1), (3, 1), (4, 1), (5, 1)\}$, illustrated in Fig. 5.19a. Then there are four arcs that have to be assigned positive numbers of lanes. The following inequality has to stay feasible $\ell(2, 1) + \ell(3, 1) + \ell(4, 1) + \ell(5, 1) \leq \mathcal{L}(1, s) + \delta - 1$ with $\delta = 4$ and $\mathcal{L}(1, s) = 3$. The value $\delta = 4$ is considered because all non-dominated patterns have to be generated and therefore all arcs in $\mathcal{A}_\alpha$ have to be assigned at least one lane. Assigning a positive number of lanes to an additional direction leading to an exit, say in this example node number 1 of the intersection, then it is always possible to assign at least one lane to it, because an additional direction increases the upper bound on the right hand side of constraint (3.32) by one.

Table 5.4 Determination of all feasible values of number of lanes

No.	1.	2.	3.	4.	5.	6.	7.	8.	9.	10.
$\ell(2,1)$	1	1	1	1	1	1	2	2	2	3
$\ell(3,1)$	1	1	1	2	2	3	1	1	2	1
$\ell(4,1)$	1	3	2	1	2	1	1	2	1	1
$\ell(5,1)$	3	1	2	2	1	1	2	1	1	1

The upper bound of lanes that can be assigned to the arcs within the intersection depends on the number of arcs with positive number of lanes and the number of lanes leaving the exit 1, i.e. $\mathcal{L}(1,s)$. For the example let $\mathcal{L}(1,s)$ be equal to 3. Then there are $\mathcal{L}(1,s) + \delta - 1 = 3 + 4 - 1 = 6$ lanes to allocate to the arcs of the subgraph. Every direction needs to be assigned at least one lane. That means we have to find all k-compositions of $n = 6$ in $k = 4$ parts. There are $\binom{n-1}{k-1} = \binom{5}{3} = 10$ possibilities, all are shown in Table 5.4. The pattern of number of lanes resulting from k-composition no. 5 in Table 5.4 is depicted in Fig. 5.19b.

References

Bretschneider S, Kimms A (2011) Pattern-based evacuation planning for urban areas. Eur J Oper Res 216:57–69

Clark J, Holton DA (1995) A first look at graph theory. World Scientific Publishing Company, Singapore

Stanley RP (1997) Enumerative combinatorics, band 1. Cambridge University Press, Cambridge

Chapter 6
A Two-Staged Heuristic Approach

In this chapter, an optimization-based heuristic approach will be presented for the one-stage pattern-based optimization model presented in Chap. 5. The heuristic approach is split into two steps: First a reduced evacuation model that disregards the detailed description of the intersections is solved. In the second step the detailed modeling of all intersections is included, but results of the first step are used such that only a subset of the feasible patterns for the intersections needs to be taken into account (compare Bretschneider and Kimms, 2011). That means the computation of the variables associated with the streets that connect intersections and the computation of the variables concerning the intersections are divided into two steps.

In Sect. 6.1 the network and the model of the first step are proposed. The network and the model of the second step are introduced in Sect. 6.2. And a computational study is presented in Sect. 6.3.

6.1 First Stage of the Heuristic

In Step 1 we work on a reduced network compared to the network used in the one-stage pattern-based evacuation model and in the second step of the heuristic approach. Intersections are modeled by one node. The goal is to find patterns and flows of arcs that are associated with streets between two intersections and an intersection and a source or sink node, such that the evacuees (units of flow) reach a safe zone (a sink) in the considered time horizon. The time horizon is subdivided into T time units, hence we consider the set of $T + 1$ points in time $\mathcal{T} = \{0, \ldots, T\}$. The reduced network is modeled with the set $\mathcal{J}$ of intersections, the set of $\mathcal{J}_D \subseteq \{d \in \mathcal{J} : \alpha_d = 1\}$ of intersections defined as sinks and the set $\mathcal{P} \subseteq \mathcal{J} \times \mathcal{J}$ of all pairs of intersections that are directly connected as follows: A source is added between two neighboring intersections i and j. It is named by the pair (i, j) (i.e. it is an element in $\mathcal{P}$, $(i, j) \in \mathcal{P}$). If one of the two intersections

S. Bretschneider, *Mathematical Models for Evacuation Planning in Urban Areas*,
Lecture Notes in Economics and Mathematical Systems 659,
DOI 10.1007/978-3-642-28759-6_6, © Springer-Verlag Berlin Heidelberg 2013

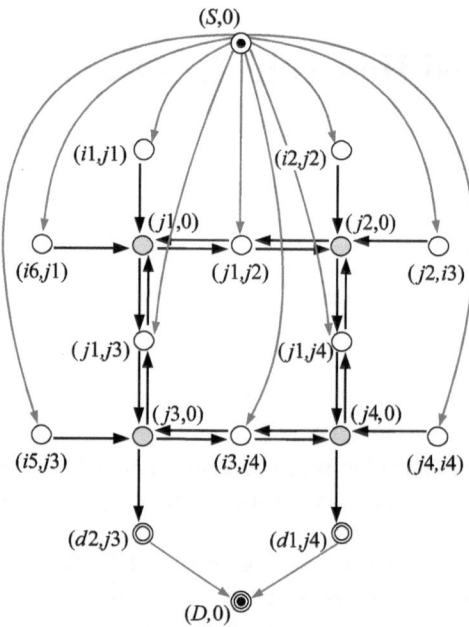

Fig. 6.1 An Illustration of the nodes and the arcs in the network $\mathbf{G}^1$ in Step 1

has only one exit, say $\alpha_i = 1$ (i.e. intersection i is a dead-end street) and $i \notin \mathcal{J}_D$, the source (i, j) is added and the intersection i is not considered further. Hence the set $\mathcal{S} = \{(i, j) \in \mathcal{P} : i, j \notin \mathcal{D}\}$ is the set of sources. Analogously, the set $\mathcal{D}$ of sink nodes is built: $\mathcal{D} = \{(i, j) \in \mathcal{P} : i \in \mathcal{D} \text{ or } j \in \mathcal{D}\}$ $(= \mathcal{P} \backslash \mathcal{S})$. Every intersection $i \in \mathcal{J}$ with $\alpha_i > 1$ is associated with a node $(i, 0)$. The set of nodes of all intersections is $\mathcal{N}_{\mathcal{J}}^1 = \{(i, 0) \in \mathcal{J} \times \{0\} : \alpha_i > 1\}$. $\mathcal{A}_C^1$ is the set of arcs that connect sources and sinks $(i, j) \in \mathcal{P}(= \mathcal{S} \cup \mathcal{D})$ with the nodes of the intersections $(i, 0), (j, 0) \in \mathcal{N}_{\mathcal{J}}^1$. The following arcs are contained in $\mathcal{A}_C^1$: Let (i, j) be an element of set $\mathcal{P}$. If $\alpha_i > 1$ and $\alpha_j > 1$ then the arcs of both directions are included in $\mathcal{A}_C^1$, i.e. $(i, j, i, 0)$, $(i, 0, i, j)$ and $(i, j, j, 0)$, $(j, 0, i, j)$. If $(i, j) \in \mathcal{D}$ and say $i \in \mathcal{J}_D$, then the arc pointing from $(j, 0)$ to the sink node (i, j) is included (if $j \in \mathcal{J}_D$ then arcs $(i, 0, i, j)$ is included). If $(i, j) \in \mathcal{S}$ and $\alpha_i = 1$, then the arc $(i, j, j, 0)$ is included. Or otherwise, if $\alpha_j = 1$, then the arc $(i, j, i, 0)$ is included in $\mathcal{A}_C^1$. Additionally, a super source $(S, 0)$ and a set $\mathcal{A}_S$ of arcs that point from the super source $(S, 0)$ at all sources in $\mathcal{S}$ as well as a super sink $(D, 0)$ and a set $\mathcal{A}_D$ that point from a sink node $(i, j) \in \mathcal{D}$ at the super sink $(D, 0)$ are added. The complete street network is represented by the network $\mathbf{G}^1 = (\mathcal{N}^1, \mathcal{A}^1)$ with $\mathcal{N}^1 = \{(S, 0)\} \cup \mathcal{S} \cup \mathcal{N}_{\mathcal{J}}^1 \cup \mathcal{D} \cup \{(D, 0)\}$ and $\mathcal{A}^1 = \mathcal{A}_S \cup \mathcal{A}_C^1 \cup \mathcal{A}_D$. The components of the network that are not subscripted by 1 are used in the network of the first step as well as in the network of the second step. Figure 6.1 illustrates a street network with the super nodes and the corresponding arcs. There are four intersections $j1, j2, j3$ and $j4$ with $\alpha = 4$ entrances/exits as well as eight intersections $i1, i2, \ldots, i6$ (dead ends) and $d1, d2$ (sinks) with $\alpha = 1$.

Let $\mathcal{PC}^1 = \bigcup_{g=1}^{G_C} \mathcal{PC}^1(g)$ be the set of all feasible non-dominated patterns of types $g = 1,\ldots,G_C$. In Step 1 the patterns are associated with the arcs A_C^1 that connect the intersections with sources and sinks. Consider two intersections i and j, where $(i,j) \in \mathcal{S}$ and $\alpha_i > 1, \alpha_j > 1$. Then a pattern includes the number of lanes of the arcs $(i,0,i,j)$, $(i,j,i,0)$, $(j,0,i,j)$ and $(i,j,j,0)$, i.e. the pattern determines the inflow and total capacities of these arcs. Now consider two intersections i and j where $(i,j) \in \mathcal{S}$ and where one of these intersections has just one entrance/exit, assume $\alpha_i = 1$, then a pattern includes the number of lanes of the arc $(i,j,j,0)$. If $(i,j) \in \mathcal{D}$ is considered, where $i \in \mathcal{J}_\mathcal{D}$, then a pattern includes the number of lanes of the arc $(j,0,i,j)$.

We need the following further parameters and sets to formulate the pattern-based optimization model in Step 1:

- $O(i,j)$ is the number of evacuees that are assigned to source $(i,j) \in \mathcal{S}$
- $\tau^1(j,l,i,h) \in \mathbb{N}$ is the travel time on arc $(j,l,i,h) \in \mathcal{A}^1$, $\tau^1(j,l,i,h) = 0$ for all $(j,l,i,h) \in \mathcal{A}_\mathcal{S} \cup \mathcal{A}_\mathcal{D}$
- $c^{in,1}(j,l,i,h,p)$ is the inflow capacity of arc $(j,l,i,h) \in \mathcal{A}_C^1$ in pattern $p \in \mathcal{PC}^1$
- $c^1(j,l,i,h,p)$ is the total capacity of the arc $(j,l,i,h) \in \mathcal{A}_C^1$ in pattern $p \in \mathcal{PC}^1$

The decision variables are the following:

- $x^1(j,l,i,h,t) \in \mathbb{R}_0^+$ is the inflow of vehicles which leave node (j,l) in time $t \in \mathcal{T}$ to reach node (i,h) in time $t + \tau^1(j,l,i,h)$ with $(j,l,i,h) \in \mathcal{A}^1, t \in \mathcal{T}$. For the ease of notation we define $x^1(j,l,i,h,t) = 0$ for $t < 0$.
- $y_C^1(i,j,p) \in \{0,1\}$: $y_C^1(i,j,p) = 1$ if pattern $p \in \mathcal{PC}^1$ is chosen for the connection $(i,j) \in \mathcal{P}$, $y_C^1(i,j,p) = 0$ otherwise.

The pattern-based evacuation model for Step 1 is formulated as follows:

Objective Function

$$\min \sum_{t \in \mathcal{T}} \sum_{(d_1,d_2) \in \mathcal{D}} \gamma(t) x^1(d_1, d_2, D, 0, t) \tag{6.1}$$

Network Flow Constraints

$$\sum_{t \in \mathcal{T}} x^1(S, 0, s_1, s_2, t) = O(i,j) \quad \text{for all } (s_1, s_2) \in \mathcal{S} \tag{6.2}$$

$$\sum_{t\in\mathcal{T}}\sum_{(d_1,d_2)\in\mathcal{D}} x^1(d_1,d_2,D,0,t) = \sum_{(s_1,s_2)\in S} O(s_1,s_2) \tag{6.3}$$

$$\sum_{(i,h)\in\mathcal{N}^1:(i,h,j,l)\in\mathcal{A}^1} x^1(i,h,j,l,t-\tau(i,h,j,l)) = \sum_{(i,h)\in\mathcal{N}^1:(j,l,i,h)\in\mathcal{A}^1} x^1(j,l,i,h,t)$$

$$\tag{6.4}$$

$$\text{for all } (j,l) \in \mathcal{N}^1\backslash\{(S,0),(D,0)\}; t \in \mathcal{T}$$

Traffic Routing Constraints

$$\sum_{p\in\mathcal{PC}^1(type(i,j))} y_C^1(i,j,p) = 1 \quad \text{for all } (i,j) \in \mathcal{P} \tag{6.5}$$

Capacity Constraints

$$x^1(i,0,i,j,t) \le \sum_{p\in\mathcal{PC}^1(type(i,j))} c^{in,1}(i,0,i,j,p)y_C^1(i,j,p) \tag{6.6}$$

$$\text{for all } (i,0,i,j) \in \mathcal{A}_C^1; t \in \mathcal{T}$$

$$x^1(i,j,i,0,t) \le \sum_{p\in\mathcal{PC}^1(type(i,j))} c^{in,1}(i,j,i,0,p)y_C^1(i,j,p) \tag{6.7}$$

$$\text{for all } (i,j,i,0) \in \mathcal{A}_C^1; t \in \mathcal{T}$$

$$\sum_{\tilde{t}\in\{t-\tau(i,0,i,j)+1,...,t\}} x^1(i,0,i,j,\tilde{t}) \le \sum_{p\in\mathcal{PC}(type(i,j))} c^1(i,0,i,j,p)y_C^1(i,j,p) \tag{6.8}$$

$$\text{for all } (i,0,i,j) \in \mathcal{A}_C^1; t \in \mathcal{T}$$

$$\sum_{\tilde{t}\in\{t-\tau(i,j,i,0)+1,...,t\}} x^1(i,j,i,0,\tilde{t}) \le \sum_{p\in\mathcal{PC}(type(i,j))} c^1(i,j,i,0,p)y_C^1(i,j,p) \tag{6.9}$$

$$\text{for all } (i,j,i,0) \in \mathcal{A}_C^1; t \in \mathcal{T}$$

The objective function (6.1) is the weighted sum of flows entering the arcs that point to the super-destination $(D, 0)$. Let the map $\gamma : \mathcal{T} \to \mathbb{R}$ be increasing. So the weights increase with the points in time, i.e. the penalty value is greater the later flow arrives in a sink. We choose $\gamma(t) = t + 1$. Due to constraint (6.2) the flow units, i.e. the evacuees, that are associated with a source have to enter the source, i.e. the street-network, within the time horizon. The assigned evacuees have to leave the corresponding source because of the flow conservation (6.4). All evacuees have to leave the evacuation zone within the considered time horizon (constraint (6.3)). Constraint (6.4) is the dynamic version of the flow conservation constraint. All flow units that enter node (j, l) in time t have to leave node (j, l) in time t. To identify the flow units that enter node (j, l) in time t, the inflow of all arcs leading to node (j, l) at time t minus the corresponding travel time of the arc has to be considered. Exactly one pattern has to be chosen for every street-segment that connects two intersections (constraint (6.5)). The constraints (6.6), (6.7) and (6.8), (6.9) restrict the inflow capacity and the total capacity on arcs for every point in time, respectively. The model is solved optimally.

6.2   Second Stage of the Heuristic

The detailed network is considered in Step 2. That means that all possible directions within intersections are included. The detailed network of Step 2 is the network of the one-stage pattern-based evacuation model (see Fig. 5.1 for an example). No crossing conflict is allowed, the number of merging conflicts of lanes within intersections is restricted, the number of usable lanes has to be regarded, and only one-way streets are considered. The results of Step 1 of the decision variables $y_C^1(i, j, p)$ are translated to the model of Step 2. They build parameters such that the directions and capacities for the street sections between intersections, sources and destinations are fixed in Step 2.

The number of lanes concerning the arcs connecting intersections is given by the results from Step 1. We can use the information to reduce the number of non-dominated intersection patterns. With the results of Step 1 the entrance/exit-nodes can be separately divided into entrances and exits of intersections and hence the number of patterns can be reduced.

We introduce for every entrance l of intersection i the value $e(i, l)$, where $e(i, l) = 0$ if (i, l) is an exit and $e(i, l) = 1$ if (i, l) is an entrance. Node (i, l) is an exit, if the chosen pattern in Step 1 allows the flow on arc $(i, l, i, j) \in \mathcal{A}_C$, where $(i, j) \in \mathcal{P}$. Node (i, l) is an entrance, if the chosen pattern in Step 1 allows the flow on arc (i, j, i, l), where $(i, j) \in \mathcal{P}$. Every intersection i has a signature $(e(i, 1), \ldots, e(i, \alpha_i))$ of entrances and exits that can be used to reduce the number of patterns. $e(i, k) = 0$ denotes that intersection node (i, k) is an exit and $e(i, k) = 1$ denotes that (i, k) is an entrance of intersection i, $k = 1, \ldots, \alpha_i$. Only these patterns that have the very same signature of entrances and exits are permitted to be chosen from. For example, Fig. 6.2 shows all intersections with

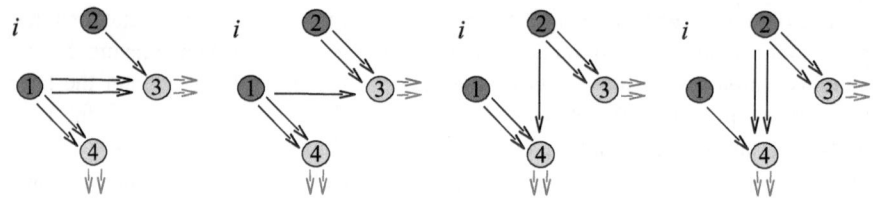

Fig. 6.2 Patterns of number of lanes of intersections with signature $(1, 1, 0, 0)$

signature $(1, 1, 0, 0)$, where the number of available lanes for every street section is two. Each arc represents a lane. Let $\mathcal{PJ}(type(i), e(i, 1), \ldots, e(i, \alpha_i))$ be the set of patterns of a signature $(e(i, 1), \ldots, e(i, \alpha_i))$ of entrances and exits, where α_i is the number of entrances/exits of the considered intersection i, $type(i) \in \{1, \ldots, G_{\mathcal{J}}\}$ is the intersection type of i and $e(i, k) \in \{0, 1\}$, $k = 1, \ldots, \alpha_i$.

New parameters and sets used Step 2 are summarized in the following:

- $\mathcal{PJ}$ is the set of all patterns for the arcs within intersections
- $y_C^1(i, j, p) \in \{0, 1\}$: $y_C^1(i, j, p) = 1$ if pattern $p \in \mathcal{PC}^1$ is chosen in Step 1 for the connection $(i, j) \in \mathcal{P}$, $y_C^1(i, j, p) = 0$ otherwise
- $\tau(j, l, i, h) \in \mathbb{N}$ is the travel time on arc $(j, l, i, h) \in \mathcal{A}$, $\tau(j, l, i, h) = 0$ for all arcs $(j, l, i, h) \in \mathcal{A}_S \cup \mathcal{A}_D$
- $c^{in}(j, l, i, h, p)$ is the inflow capacity of arc $(j, l, i, h) \in \mathcal{A}_C$ in pattern $p \in \mathcal{PC}$ and $(j, l, i, h) \in \mathcal{A}_{\mathcal{J}}$ in pattern $p \in \mathcal{PJ}$
- $c(j, l, i, h, p)$ is the total capacity of arc $(j, l, i, h) \in \mathcal{A}_C$ in pattern $p \in \mathcal{PC}$ and $(j, l, i, h) \in \mathcal{A}_{\mathcal{J}}$ in pattern $p \in \mathcal{PJ}$

Note, the counterpart in Step 1 of the "second-step" arc $(i, j, i, l) \in \mathcal{A}_C$ with $(i, j) \in \mathcal{S}$ is the "first-step" arc $(i, j, i, 0) \in \mathcal{A}_C^1$ and the counterpart of the "second-step" arc $(i, l, i, j) \in \mathcal{A}_C$ with $(i, j) \in \mathcal{S}$ is the "first-step" arc $(i, 0, i, j) \in \mathcal{A}_C^1$. Therefore we set $c^{in}(i, j, i, l, p) = c^{in,1}(i, j, i, 0, p)$ and $c^{in}(i, l, i, j, p) = c^{in,1}(i, 0, i, j, p)$, respectively. Analogously, the parameters total capacity $c(j, h, i, l, p)$ and travel time $\tau(j, h, i, l, p)$, $(j, h, i, l) \in \mathcal{A}_C$ are set for the second step. In the same way the patterns $\mathcal{PC}^1$ of the first step are translated to the pattern $\mathcal{PC}$ of the second step, the corresponding capacity parameters are associated with the counterparts of the arcs in Step 2 (as described above).

The decision variables of Step 2 are:

- $x(j, l, i, h, t) \in \mathbb{R}_0^+$ is the inflow variable that is defined analogously to the inflow variable x^1 in Step 1, for all $(j, l, i, h) \in \mathcal{A}$, $t \in \mathcal{T}$, and it is set $x(j, l, i, h, t) = 0$ for all $t < 0$
- $y_{\mathcal{J}}(j, p) \in \{0, 1\}$: $y_{\mathcal{J}}(j, p) = 1$ if pattern $p \in \mathcal{PJ}$ is chosen for intersection $j \in \mathcal{J}$, $y_{\mathcal{J}}(j, p) = 0$ otherwise

The model in Step 2 is formulated as follows:

Objective Function

$$\min \sum_{t \in T} \sum_{(d_1, d_2) \in \mathcal{D}} \gamma(t) x(d_1, d_2, D, 0) \tag{6.10}$$

Network Flow Constraints

$$\sum_{t \in T} x(S, 0, s_1, s_2, t) = O(s_1, s_2) \quad \text{for all } (s_1, s_2) \in \mathcal{S} \tag{6.11}$$

$$\sum_{t \in T} \sum_{(d_1, d_2) \in \mathcal{D}} x(d_1, d_2, D, 0, t) = \sum_{(s_1, s_2) \in S} O(s_1, s_2) \tag{6.12}$$

$$\sum_{(i,h) \in \mathcal{N}:(i,h,j,l) \in \mathcal{A}} x(i, h, j, l, t - \tau(i, h, j, l)) = \sum_{(i,h) \in \mathcal{N}:(j,l,i,h) \in \mathcal{A}} x(j, l, i, h, t) \tag{6.13}$$

$$\text{for all } (j, l) \in \mathcal{N} \setminus \{(S, 0), (D, 0)\}; t \in T$$

Traffic Routing Constraints

$$\sum_{p \in \mathcal{PJ}(type(i), e(i,1), \dots, e(i, \alpha_i))} y_{\mathcal{J}}(i, p) = 1 \quad \text{for all } i \in \mathcal{J} \tag{6.14}$$

Capacity Constraints

$$x(i, l, i, h, t) \leq \sum_{p \in \mathcal{PJ}(type(i), e(1), \dots, e(\alpha_i))} c^{in}(i, l, i, h, p) y_{\mathcal{J}}(i, p) \tag{6.15}$$

$$\text{for all } (i, l, i, h) \in \mathcal{A}_{\mathcal{J}}; t \in T$$

$$\sum_{\tilde{t} \in \{t - \tau(i,l,i,h) + 1, \dots, t\}} x(i, l, i, h, \tilde{t}) \leq \sum_{p \in \mathcal{P}_{\mathcal{J}}(type(i), e(1), \dots, e(\alpha_i))} c(i, l, i, h, p) y_{\mathcal{J}}(i, p) \tag{6.16}$$

$$\text{for all } (i, l, i, h) \in \mathcal{A}_{\mathcal{J}}; t \in T$$

$$x(i,l,i,j,t) \leq \sum_{p \in \mathcal{PC}(type(i,j))} c^{in}(i,l,i,j,p) y_C^1(i,j,p) \qquad (6.17)$$

for all $(i,l,i,j) \in \mathcal{A}_C$ with $(i,l) \in \mathcal{N}_{\mathcal{J}}; t \in \mathcal{T}$

$$x(i,j,i,l,t) \leq \sum_{p \in \mathcal{PC}(type(i,j))} c^{in}(i,j,i,l,p) y_C^1(i,j,p) \qquad (6.18)$$

for all $(i,j,i,l) \in \mathcal{A}_C$ with $(i,l) \in \mathcal{N}_{\mathcal{J}}; t \in \mathcal{T}$

$$\sum_{\tilde{t} \in \{t - \tau(i,l,i,j)+1...t\}} x(i,l,i,j,\tilde{t}) \leq \sum_{p \in \mathcal{PC}(type(i,j))} c(i,l,i,j,p) y_C^1(i,j,p) \qquad (6.19)$$

for all $(i,l,i,j) \in \mathcal{A}_C$ with $(i,l) \in \mathcal{N}_{\mathcal{J}}; t \in \mathcal{T}$

$$\sum_{\tilde{t} \in \{t - \tau(i,j,i,0)+1...t\}} x(i,j,i,l,\tilde{t}) \leq \sum_{p \in \mathcal{PC}(type(i,j))} c(i,j,i,l,p) y_C^1(i,j,p) \qquad (6.20)$$

for all $(i,j,i,l) \in \mathcal{A}_C$ with $(i,l) \in \mathcal{N}_{\mathcal{J}}; t \in \mathcal{T}$

The objective function (6.10) and the constraints (6.11)–(6.13) are equivalent to the objective (6.1) and the constraints (6.2)–(6.4) in Step 1. Exactly one pattern has to be chosen for every intersection $i \in \mathcal{J}$ (constraint (6.14)). The constraints (6.15) and (6.16) restrict the inflow capacity and the total capacity on arcs within intersections at every point in time, respectively. The constraints (6.17), (6.18) and (6.19), (6.20) restrict the inflow capacity and the total capacity on arcs in $\mathcal{A}_C$, respectively. They are equivalent to the constraints (6.6)–(6.9) in Step 1, but in Step 2 the decision variables $y_C^1(i,j,p)$ from Step 1 are parameters in Step 2 obtained in Step 1. We solve this model optimally.

6.3   Computational Study

All optimization models and heuristics, the first- and second-stage models and the relaxations of the one-stage evacuation model, the two-staged solution approach as well as the relaxation-based approach were implemented with AMPL using CPLEX 10.0 and were run on a personal computer with an AMD Athlon(tm) 64X2 Dual Core Processor 4600+ (2.41 GHz) with 1.96 GB RAM. The computational study is taken from Bretschneider and Kimms (2011).

6.3.1 Sets of Instances

The considered 1, 180 instances were based on grid networks of different sizes with different exit-patterns and numbers of inhabitants. An $M \times N$ grid network was generated as a network with $M \cdot N$ intersections with four exits/entrances and $2(M + N)$ intersections with one entrance or exit at the boundary. The latter intersections were assumed to be the destinations or sources on the boundary of the considered network. Figure 6.3 illustrates such a grid network of the size $M \times N$. The intersections with four exits/entrances are labeled with $j1, \ldots, jMN$ (j stands for junction).

Two groups of instances were considered. They were varied in the network size, the numbers of evacuees and the capacity.

The parameters of the first group of instances were chosen as follows:

- Five grid sizes: 3×4, 4×5, 5×5, 5×6 and 5×7
- Two exit-patterns: the first exit-pattern denotes that the all boundary nodes are exits, while the second exit-pattern denotes that the right and bottom boundary nodes are exits, see Fig. 6.3
- Ten different patterns of the numbers of evacuees randomly chosen from each of the following three sets $\{0, 50, 100\}$, $\{0, 20, 40, 60\}$ and $\{0, 15, 30, 45, 60, 75\}$
- The number of usable lanes for all street segments was set to 2, 3 or 4
- Travel time τ: $\tau = 7$ for all arcs $\mathcal{A}_C^1$ in Step 1 and for all arcs in $\mathcal{A}_C$ in Step 2, $\tau = 1$ for all arcs in $\mathcal{A}_{\mathcal{J}}$ in Step 2
- Inflow capacity c^{in} and total capacity c per lane: the inflow capacity of every lane was set to 1, the total capacity was set to 8 for lanes of arcs in $\mathcal{A}_C^1$ in Step 1 as well as for all arcs in $\mathcal{A}_C$ in Step 2, moreover the total capacity was set to 1 for lanes of arcs in $\mathcal{A}_{\mathcal{J}}$ in Step 2
- The time horizon was subdivided into $T = 200$ time units

The parameters of the second group of instances were chosen as follows:

- Seven grid sizes: 6×8, 7×8, 7×9, 8×9, 7×10, 8×10 and 9×10
- One exit-pattern: the second exit-pattern from above, i.e. it denotes that the right and bottom boundary nodes are exits, see Fig. 6.3
- Ten different patterns of numbers of evacuees randomly chosen from each of the following two sets $\{0, 10, 20\}$ and $\{0, 5, 10, 15\}$
- The number of usable lanes for all street segments was set to 3 or 4
- Travel time τ: $\tau = 7$ for all arcs $\mathcal{A}_C^1$ in Step 1 and for all arcs in $\mathcal{A}_C$ in Step 2, $\tau = 1$ for all arcs in $\mathcal{A}_{\mathcal{J}}$ in Step 2
- Inflow capacity c^{in} and total capacity c per lane: the inflow capacity of every lane was set to 1, the total capacity was set to 5 for lanes of arcs in $\mathcal{A}_C^1$ in Step 1

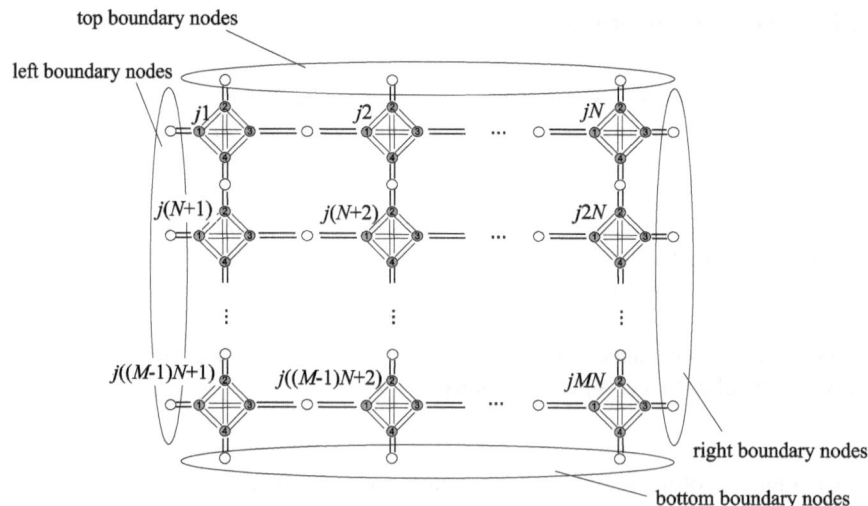

Fig. 6.3 An $M \times N$ grid network

as well as for all arcs in $\mathcal{A}_C$ in Step 2, moreover the total capacity was set to 1 for lanes of arcs in $\mathcal{A}_{\mathcal{J}}$ in Step 2
• The time horizon was subdivided into $T = 200$ time units

In total, $1,180 (= 900 + 280)$ instances were generated.

6.3.2   Models for Comparison

Percentage gaps were computed to evaluate the objective function values achieved with the two-staged heuristic approach. We chose relaxations of the one-stage pattern-based evacuation model and as well as the relaxation-based approach (see Chap. 4) to generate values for comparison.

Relaxations of the One-Stage Pattern-Based Evacuation Model

The percentage gap between the result of the two-staged approach and the result of the corresponding relaxation of the one-stage evacuation model was computed with

$$\% - \mathrm{GAP}_1 = \frac{UB_1 - LB}{LB} \cdot 100,$$

where UB_1 denotes the objective function value of the feasible solution achieved with the two-staged heuristic approach, LB denotes the objective function value of the optimal solution of the corresponding relaxation of the instances.

The relaxation chosen for the instances with network sizes 3×4, 4×5, 5×5, 5×6 and 5×7 is the LP-relaxation of the one-stage pattern-based evacuation model. To compare the objective function values of the remaining instances we chose additional relaxations of the one-stage pattern-based model because the LP-relaxation of these instances could not be solved on the used computer. An out-of-memory error occurred. In addition to the relaxations, the number T of time units was decreased individually for the set of instances of every network size. The reduction depended on the results of the two-staged approach of the corresponding instances. We set T for the instances with the network size $M \times N$ to the maximum evacuation time of the evacuation times resulting from the two-staged approach of the different instances with the network size $M \times N$. The relaxation that we chose for the instances with the network sizes 6×8, 7×8 and 7×9 is the relaxation of constraint (6.14) from "exactly one pattern has to be chosen" to "two or less patterns have to be chosen" (i.e. from "$= 1$" to "≤ 2"). We chose for the instances with network sizes 8×9, 7×10, 8×10 and 9×10 the following relaxed LP-relaxation: the patterns are neglected but the capacities depending on the number of provided lanes of every arc are still taken into account. That means every arc can be traveled by flow without restrictions except the capacity restrictions.

Relaxation-Based Heuristic Approach for an MIP-Evacuation Model

The urban one-way evacuation model presented in Sect. 3.4 with $\gamma(t) = t + 1$ is considered and the instances are solved with the relaxation-based heuristic approach adjusted for the one-way case (see Sect. 4.2).

The percentage gap between the result of the relaxation-based approach and the corresponding result of the relaxation is computed as follows

$$\% - \text{GAP}_2 = \frac{UB_2 - LB}{LB} \cdot 100,$$

where UB_2 denotes the objective function value of the feasible solution achieved with the relaxation-based approach and LB denotes the objective function value of the optimal solution of the corresponding result of the relaxations of the one-stage model of the instances.

6.3.3 Computational Results

The total number of investigated instances was $1,180$ but in a few cases no feasible integer solution was found in Step 1 of the pattern-based heuristic approach within the given maximal computation time of $3 \, \text{h}$. The time limit was set for the 160 instances with networks of the sizes 8×9, 7×10, 8×10 and 9×10. In 27 of 160 times the time limit was reached but just six times no integer solution was found.

Table 6.1 Average percentage gaps $\% - GAP_1$ and $\% - GAP_2$

Size	Exit-pattern	Avg $\% - GAP_1$	Avg $\% - GAP_2$
3×4	First	0.42	4.81
4×5	First	0.95	6.45
5×5	First	1.33	7.15
5×6	First	1.62	7.95
5×7	First	1.76	9.10
3×4	Second	2.71	11.90
4×5	Second	2.76	14.06
5×5	Second	2.88	13.15
5×6	Second	2.95	13.46
5×7	Second	3.14 (3.30)	14.70
6×8	Second	1.00	1.47
7×8	Second	0.96	1.82
7×9	Second	0.92	1.13
8×9	Second	0.89 (0.91)	0.97
7×10	Second	0.68 (0.631)	2.60
8×10	Second	0.72	–
9×10	Second	0.93	–

That is one of the instances based on network 7×10 and five of the instances based on network 8×10. The time limit was reached with integer solutions within the set of instances of the networks 7×10, 8×10, and 9×10. The six instances with no feasible solution are not considered within the average gap calculation in Table 6.1, but the computation times of all investigated $1,180$ instances are considered in the calculation of the average computation time in Table 6.3.

The computation of a feasible solution with the relaxation-based approach was not always possible within the given time limit of 6 h. The computation stopped in 16 times before a feasible solution could be found and the computation stopped in 95 times without solution because the computations of the instances required too much memory. Hence, the relaxations that have to be computed in relaxation-based heuristic approach could not be computed in 111 of $1,180$ cases. That are all instances based on networks 8×10 and 9×10, nearly half the number of instances based on network 8×9 (17 of 40), and a few of the instances based on networks 7×10 (4 of 40) and 5×7 with the second exit-pattern (10 of 90). I.e. in total there are $1,069$ of the $1,180$ instances that can be compared to the two-staged approach. The average percentage gaps $\% - GAP_1$ within the brackets in the row of the networks 5×7 second, 8×9, and 7×10 in Table 6.1 show the average percentage gaps over the instances that have a feasible result in the relaxation-based approach and can be compared directly with the associated percentage gap $\% - GAP_2$. The computation times of all $1,180$ investigated instances are considered in Table 6.3.

The resulting percentage gaps $\% - GAP_1$ and $\% - GAP_2$ were aggregated over a fixed network size and a fixed exit-pattern. They are shown in Table 6.1.

The average percentage gaps $\% - GAP_1$ that compare the results of the pattern-based heuristic approach with the results of the relaxations are small. The largest percentage gap $\%-GAP_1$ occurs in the instances of the network 4×5 with the second exit-pattern; it is 7.36%. The value $\% - GAP_2$ shows the percentage gap of the upper bound achieved with the relaxation-based heuristic approach and the lower bound computed with the corresponding relaxation of the one-stage model. The average gaps $\% - GAP_2$ for the networks are always greater than the corresponding average percentage gaps $\% - GAP_1$. But the value of $\% - GAP_2$ is not greater than the value of $\% - GAP_1$ for every instance: in 73 of 1,069 cases the relaxation-based approach generates solutions with better objective function values than the ones from the second-stage approach. In the first group of instances the relaxation-based approach was better than the pattern-based approach in 9 of 890 instances, i.e. in 1.01% of the considered instances. In the second group of instances the relaxation-based approach was cases better in 64 of 179, i.e. 35.75% of the considered instances. Besides the network size the difference between these instances is the number of evacuees that have to leave the evacuation area and hence how "crowded" the street-networks of the different instances are. The more "crowed", the better the two-staged approach seems to be in comparison to the relaxation-based approach.

A value we could consider is the average number of evacuees per source of the instances based on a network. We calculated the average number of evacuees per source as the average number of evacuees of the instances based on a certain network divided by the number of sources. The number of sources stays the same for every instance based on the same network. The following observation could be made with the value of the average number of evacuees of a source: The average percentage gaps $\% - GAP_1$ and $\% - GAP_2$ tend to be greater the greater the average number of evacuees per source is. The instances with a greater average number of evacuees per source seem to be more difficult to solve for the presented heuristic approaches. The average number of evacuees per source varies from approximately 22–28 for the first group of instances with the first exit-pattern, for the first group of instances with the second exit-pattern it is approximately 40, and for the second group of instances it is approximately 9.

To compare the estimated evacuation times of the results of the pattern-based and relaxation-based heuristic approaches, the percentage gap between the estimated evacuation times of the results generated by the heuristics were computed by

$$\% - GAP = \frac{(EET_2 - EET_1)}{EET_2} \cdot 100,$$

where EET_1 is the estimated evacuation time resulting from the solution of an instance achieved by the pattern-based two-staged approach and EET_2 denotes the estimated evacuation time resulting from the solution of the same instance computed with the relaxation-based approach. Let the point in time when the last cumulated total inflow reaching the sinks that is greater than 0.05 be the estimated evacuation time. That means, if $\tilde{t} \in \mathcal{T}$ is the last point in time with $\sum_{(d_1,d_2)\in\mathcal{D}} x(d_1, d_2, D, 0, \tilde{t}) > 0.05$, then $\tilde{t}$ is the estimated evacuation time of the

Table 6.2 Average percentage gaps between the estimated evacuation times of the two-staged approach and the LP-relaxation and the number of instances where the evacuation time of the relaxation-based approach is less than the evacuation time of the pattern-based approach

Size	Exit-pattern	Avg % − GAP	# $(EET_1 > EET_2)$
3×4	1st	13.26	3
4×5	1st	14.82	3
5×5	1st	16.11	1
5×6	1st	15.67	0
5×7	1st	19.52	0
3×4	2nd	17.24	1
4×5	2nd	17.74	0
5×5	2nd	16.02	0
5×6	2nd	15.43	0
5×7	2nd	16.94	0
6×8	2nd	1.47	5
7×8	2nd	2.26	0
7×9	2nd	0.16	6
8×9	2nd	0.71	0
7×10	2nd	0.93	2

solution of the considered instance (i.e. $\sum_{(d_1,d_2)\in\mathcal{D}} x(d_1, d_2, D, 0, t) \leq 0.05$ for $t > \tilde{t}$).

The percentage gaps % − GAP were aggregated over a fixed network size and a fixed exit-pattern. The average percentage gaps % − GAP and the number of instances where the estimated evacuation time of the solution of the pattern-based approach is greater than the estimated evacuation time of the solution of the relaxation-based approach (# $(EET_1 > EET_2)$) are shown in Table 6.2.

The pattern-based heuristic approach leads almost always to a faster estimated evacuation time than the relaxation-based heuristic approach; in 21 of 1,096 cases the estimated evacuation time of the result computed with the relaxation-based approach is better than the estimated evacuation time of the result generated by the pattern-based approach, i.e. in ca. 2% of the instances. The greatest gap % − GAP of 55.71% (60 time units/149 time units) occurs in one of the instances based on network 3×4 with the second exit-pattern. There is at least one instance based on each of the networks with the first exit-pattern that has a percentage gap % − GAP of approximately 50%. The percentage gap % − GAP of the estimated evacuation times in one of the instances based on network 4×5 (second exit-pattern) is greater than 50%. The greatest percentage gaps % − GAP of instances based on networks 5×5, 5×6 and 5×7 with the second exit-pattern are approximately 35%. The average of the biggest gaps of the instances based on the networks 6×8 through 7×10 is 13.68%.

The smallest gap % − GAP occurs for an instance based on the network 5×5 (first exit-pattern) with −5.13%, i.e. here the relaxation-based approach leads to a better estimated evacuation time.

Table 6.3 Average computation times of the approaches and the relaxation in CPU-seconds

Size	Exit-pattern	Avg run-time 2 stages	Avg run-time relaxation	Avg run-time relax.-based appr.
3×4	First	16	275	99
4×5	First	147	797	386
5×5	First	195	1,318	740
5×6	First	655	1,919	1,184
5×7	First	1,189	2,522	1,616
3×4	Second	30	336	111
4×5	Second	174	973	1,016
5×5	Second	605	1,620	2,259
5×6	Second	1,406	2,222	4,341
5×7	Second	2,539	2,900	(5,783)
6×8	Second	180	574	3,177
7×8	Second	298	1,109	4,533
7×9	Second	316	1,276	4,656
8×9	Second	418	222	(4,209)
7×10	Second	6,281	462	(11,277)
8×10	Second	5,460	582	(778)
9×10	Second	3,110	632	(816)

The average computation times of the two-staged approach, of the relaxations of the one-stage model and the computation times of the relaxation-based approach are shown in Table 6.3. The computation times of all instances are included, even the computation times of the instances with the result "out of memory" (e.g. all instances based on networks 8×10 and 9×10) or with the result that no feasible solution could be found within the given time limit (e.g. a few instances based on networks 5×7, 8×9 and 7×10). Those average computation times that include solutions with no feasible result are labeled with brackets.

The average computation times of the two-staged approach are always shorter than the average computation times of the LP-relaxation and the relaxed LP-relaxation of the instances with the network sizes 3×4 through 7×9. The number of instances that are based on the networks with sizes 3×4 through 7×9 is $1,020$. The computation times of the two-staged approach are almost always shorter than the computation times of the corresponding relaxation in these cases. In 79 of the $1,020$ instances the computation time of the two-staged solution approach took longer than the corresponding relaxation. The relaxation used to compare with the instances based on networks of the size 8×9 through 9×10 is a much easier model than all the other ones. So the solution times of this relaxation are unsurprisingly shorter than all the corresponding instances of the other models.

The average computation times needed to generate a solution with the relaxation-based approach are always longer than the average computation times of the two-staged approach for the instances of the networks 3×4 through 7×10. In 95 cases out of $1,069$, i.e. just in 8.89%, the computation time of the two-staged approach took longer than the computation time of the relaxation-based approach. The average computation times required by the relaxation-based approach of the instances of the networks 8×10 and 9×10 are the average computation times until the error message concerning the memory capacity arises. No feasible solution of the 80 instances based on these networks could be generated with the relaxation-based approach. The average computation times of these instances are short compared to the computation times needed to generate a feasible solution with the two-staged heuristic approach. But with the two-staged heuristic approach, a feasible solution could be generated for 75 of 80 instances that are based on networks 8×10 and 9×10.

Reference

Bretschneider S, Kimms A (2011) Pattern-based evacuation planning for urban areas. Eur J Oper Res 216:57–69

Chapter 7
A Multicommodity Urban Evacuation Problem

The multicommodity problem that will be introduced in this chapter is formulated as a mixed-integer model where the number of lanes are represented by integer variables. It incorporates the commodities vehicles of evacuees and buses. The two-staged heuristic approach solving the pattern-based formulation of the urban evacuation problem turns out to generate feasible solutions faster than the relaxation-based heuristic approach that solves the mixed-integer formulation of the urban evacuation problem. With the two-staged approach it is possible to generate feasible solutions of problem instances which cannot be solved with the relaxationbased approach. The computational study comparing both approaches is restricted on one-way streets (see Sect. 6.3). But for the multicommodity problem, we want to allow two-way streets. One reason is that buses, which drive to assembly areas to pick up evacuees, travel opposite to the evacuation flow.

When permitting two-way streets, the number of patterns increases significantly. The number of not unreasonable (and non-dominated) patterns of street sections between neighboring intersections increases from 3 (one-way case, see Sect. 5.2) to 8, 15 and 24 (two-way case) for street sections with 2, 3 and 4 usable number of lanes, respectively. Nodes of intersections may be entrances and exits at the same time. Hence, we decided to formulate the multicommodity problem as a mixed-integer model. However, the heuristic approach to find a feasible solution for the multicommodity problem pursues a similar strategy as the two-staged heuristic approach (see Chap. 8): certain steps operate on a reduced network (the intersection of a street network are represented by a single node) and in following steps details of intersections are taken into account while incorporating result from preceding steps.

7.1 Introduction and Assumptions

In the case of an emergency evacuation it is unlikely that all evacuees can reach the safe zone without help. Not everybody has an own car and not everybody can be given a lift. For these evacuees buses can be provided at assembly areas like schools, parking places or other public places. In cases of hospitals or nursing homes it is likely that ambulances are needed.

Buses could transport the evacuees from an assembly point to a safe place outside the evacuation area, unload the evacuees and travel back to an assembly point to load passengers again if necessary. Buses could be already at the assembly points or they have to travel from the area outside the evacuation zone to an assembly point. In the latter case the buses drive in the opposite direction of the flow of evacuees leaving by their vehicles or the provided buses the evacuation zone. Ambulances can be handled the same way as buses: they can be modeled like buses with just one place for an evacuee that needs special medical attendance. Ambulances may travel from a special place (a hospital or a nursing home) within the evacuation area to a safe zone and back.

The flow of buses traveling from a safe place into the evacuation area, i.e. against the evacuation flow, results to new problems: If still all crossing conflicts were prohibited, feasibility could not be assured in every case. Therefore the constraints of prohibiting crossing conflicts have to be relaxed. Figure 7.1 shows an example of a small street network where a crossing conflict is unavoidable. Let the street network has a single exit to a safe area where buses can start to enter the evacuation zone and an assembly point that is the intermediate destination of the buses to pick up evacuees. The dotted line shows the route of the bus from the safe area to the assembly point. The route of vehicles of the evacuees represented by the solid line cannot be avoided. All evacuees must have the possibility to leave the evacuation area. In the illustration in Fig. 7.1, a crossing conflict within the route of vehicles from a starting point inside the evacuation zone to the safe area occurs and cannot be avoided.

For that reason two different lane types are considered: Lanes are partitioned into public- and emergency-lanes but only within intersections. On street segments between intersections only public-lanes are allowed. Public-lanes can be used by all commodities, i.e. vehicles of evacuees as well as emergency vehicles. Whereas emergency-lanes can only be traveled by emergency vehicles, like buses or ambulances. The following rules have to be obeyed:

- Crossing conflicts of two public-lanes within intersections are prohibited.
- Crossing conflicts of emergency-lanes with emergency- or public-lanes within intersections are allowed, but if an emergency-lane is needed (i.e. at least one crossing conflict occurs) the assignment of the emergency-lane is penalized.
- Merging conflicts are restricted as usual. Within the intersection both lane-types are considered and counted.

An argumentation for permitting crossings with emergency-lanes is that drivers of emergency vehicles and buses have much more experience and/or are more trained than a "normal" person. In the following we restrict on buses as emergency vehicles. Ambulances can be modeled analogously, see the discussion above.

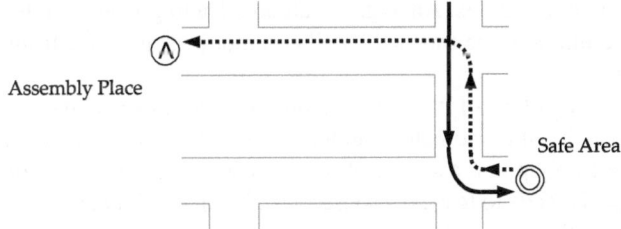

Fig. 7.1 Examples of unavoidable crossing conflicts

Commodities

The commodities that are considered in the developed model are vehicles of evacuees and buses. Evacuees that have the chance to leave the evacuation zone by car are called vehicles in the following. Vehicles can reach the safe zones without help. They do not enter or reenter the evacuation zone; they just leave the evacuation zone.

Starting at assembly places that are known in advance and fixed, evacuees that cannot reach the safe zones on their own have the possibility to go by bus. These evacuees, pedestrians, have to go to assembly points. But pedestrians are neither modeled directly nor are the flows of pedestrians considered. They are considered indirectly by the number of buses. A number of pedestrians is assigned to every assembly point in advance and with these values the number of buses that are needed to rescue all pedestrians at the assembly point can be calculated.

Buses can enter, leave and reenter the evacuation zone. The commodity bus is split into two commodities: the commodity bus_{in} and bus_{out}.

The buses bus_{in} drive from safe places outside the evacuation area without stopover to their destination, an assembly point within the evacuation area where the evacuees wait. There they "change" into the commodity bus_{out} and are ready to transport the evacuees to a safe zone. Buses bus_{out} are buses that transport the evacuees from an assembly point inside the evacuation area without stopover to a safe place outside the evacuation area. There are now sources outside the evacuation area where the flow of commodity bus_{in} starts and sinks within the evacuation area where bus_{in} ends. An assumption is that the sources of bus_{in} may be used as destinations of bus_{out} and that the sources of bus_{out} may be the destinations of bus_{in} and vice versa, respectively. That means $S(bus_{in}) = D(bus_{out})$ and $S(bus_{out}) = D(bus_{in})$, where $S(co)$ and $D(co)$ denote the set of sources of commodity co and the set of sinks of commodity co ($co \in \{bus_{in}, bus_{out}\}$), respectively. Otherwise, we have to model different constraints for nodes that are sources and sinks in the same time and nodes that are either sources or sinks (see the comments after constraints (7.11) and (7.12)). Different commodities may have different sources and sinks. Hence it depends on the commodity and their predetermined sources, weather an arc that leads from the super source S to a source is defined for flow of a certain commodity. Not all arcs that lead from S to a node are permitted to be used by all commodities. Analog, not all arcs leading from a node to the super sink D are permitted to be traveled by all commodities and vary from commodity to commodity.

The number $Bout(s)$ of buses bus_{out} that have to leave source $s \in S(bus_{out})$ are determined by the estimated number of evacuees that are assigned to the associated assembly point, a source of bus_{out}. The number $O(co, s)$ of buses $co \in \{bus_{in}, bus_{out}\}$ is ready to leave the source $s \in S(co)$ at point in time 0. The number of buses bus_{out} that are still needed to transport all evacuees to a safe zone can easily be determined with the value $O(bus_{out}, s)$, that is $Bout(s) - O(bus_{out}, s)$ for source $s \in S(bus_{out})$. That means $Bout(a) - O(bus_{out}, a)$ buses of commodity bus_{in} have to reach assembly area $a \in S(bus_{out})(= D(bus_{in}))$. The bus commodities have to follow the inflow and total capacity constraints as well as the network flow constraints. Additional, a capacity on the number of buses that are ready to start from the sources at each point in time is introduced.

In general, the travel times for arcs leading from or to a super node are set to 0. Here in the case of the bus commodities these travel times can be used to represent the time needed in assembly or safe areas after the arrival to make the buses ready to start again. At the assembly points that could be the time the pedestrians need to board the bus or at the safe places the time needed to empty the bus and/or drive to a shelter and back to an entrance to the evacuation zone (i.e. the entrance to the evacuation zone is a sink of bus_{out} and a source of bus_{in}).

The delays in the safe zone are represented with the travel times of the buses bus_{out} leaving the evacuation area, i.e. $\tau(bus_{out}, a, D) \geq 0$. Then the travel time of the buses entering the evacuation zone can be set to 0, i.e. $\tau(bus_{in}, S, a) = 0$. That means a bus that have left the evacuation zone needs time to get ready to restart again. An arriving bus at a safe zone has to be emptied and buses that are located at the beginning of the time horizon are ready to enter the evacuation zone without delay. On the other hand with $\tau(bus_{in}, S, a) > 0$ a delay could be modeled for the buses to represent the travel time needed to leave their starting place to the entrance $a \in S(bus_{in})$ to the evacuation zone.

The delays at assembly points are represented by the travel times of the buses bus_{out} assigned to arcs pointing from the super source S to a assembly node $a \in S(bus_{out})$, i.e. $\tau(bus_{out}, S, a) \geq 0$. Then the travel time of the buses bus_{in} entering the assembly points can be set to 0, i.e. $\tau(bus_{in}, a, D) = 0$. A leaving bus has to be filled.

The commodity of vehicles has the same properties as the vehicles in the single-commodity case. Vehicles have to leave the evacuation area within the given time horizon. They start in sources $s \in S(veh)$ in the evacuation zone and have to reach a sink $d \in D(veh)$ outside the evacuation zone. Flow constraints, total and inflow capacity constraints have to be respected.

Figure 7.2 illustrates the street network and the arcs that are permitted to be used by the different commodities. From the super source S the flow of commodities is distributed to the corresponding starting nodes, the sources. All commodities start there, the flows of vehicles veh and bus_{out} that start inside the evacuation zone and the flows bus_{out} that start outside the evacuation zone. The super sink D has an

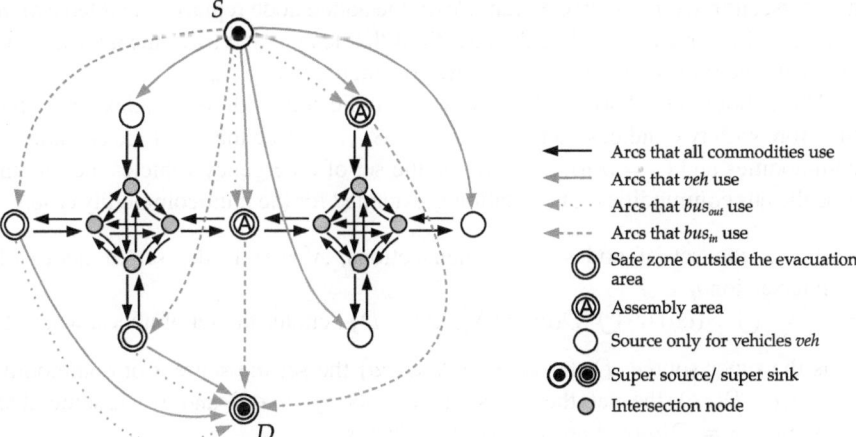

Fig. 7.2 Illustration of the network of the multicommodity model

analog property, it is a virtual destination of all commodities. The gray arcs denote the arcs that can be used only by the corresponding commodity. The buses bus_{in} start from the super source S to a node outside the evacuation area (dashed gray arcs) to travel through the street network to the corresponding destination (black arcs), an assembly node that is indicated with an **A**, and travels finally to the super sink D (a dashed gray arc). The buses bus_{out} start from the super source S to an assembly point (dotted gray arcs) to travel through the street network to the corresponding destination (black arcs), a safe zone, and travels finally to the super sink D (a dotted gray arc). The arcs only defined for veh are solid gray.

7.2   A Multicommodity Evacuation Model

7.2.1   A Network Representation

The network of the single-commodity models is extended by the sources and sinks of the commodities and the appropriate arcs leaving the super source and leading to the super sink. That means the network $\mathbf{G} = (\mathcal{N}, \mathcal{A})$ for the multicommodity model consists of the street network, artificial super nodes and the corresponding artificial arcs pointing from the super source to the sources of the commodities and pointing from the sinks to the super sink. The inflow of commodities is defined on these arcs adjacent to the super nodes only if the node is associated to the commodity as source or sink.

Again, the street network is modeled in detail with every entrance/exit of every intersection denoted by intersection nodes $\mathcal{N}_{\mathcal{J}}$ and every possible direction within the intersection denoted by the intersection arcs $\mathcal{A}_{\mathcal{J}}$. Between every pair of neighboring intersections a node is added. The set of arcs connecting intersections and added nodes is denoted by $\mathcal{A}_C$. If a pair of neighboring intersections contains an intersection with only one entrance/exit, the added node replaces the intersection with $\alpha = 1$ entrance/exit. Here the directions that lead to and leave the node associated with the intersection with $\alpha = 1$ are considered in any case.

Throughout the chapter, the nodes are indicated with one index and arcs therefore with two indices. Let $C = \{bus_{in}, bus_{out}, veh\}$ be the set of the considered commodities and $C_e = \{bus_{in}, bus_{out}\}$ be the set of emergency vehicles, i.e. in our case the bus commodities. The notation is extended for the multicommodity case:

- $\mathcal{N}_{\mathcal{J}}$ is the set of nodes of all intersections, $\mathcal{N}_{\mathcal{J}}(q)$ is the set of nodes of intersection $q \in \mathcal{J}$
- $\mathcal{N} = \bigcup_{co \in C} \mathcal{S}(co) \cup \bigcup_{co \in C} \mathcal{D}(co) \cup \mathcal{N}_{\mathcal{J}} \cup \{S, D\}$ denotes the set of nodes, where S is the super source, D the super sink, $\mathcal{S}(co)$ the set the sources of commodity $co \in C$, $\mathcal{D}(co)$ the set the sinks of commodity $co \in C$ and we assume that $\mathcal{S}(bus_{out}) = \mathcal{D}(bus_{in})$ and $\mathcal{S}(bus_{in}) = \mathcal{D}(bus_{out})$
- $\mathcal{A}_{\mathcal{J}}$ denotes set of arcs of all intersections, $\mathcal{A}_{\mathcal{J}}(q)$ denotes set of arcs of intersection $q \in \mathcal{J}$
- $\mathcal{A}_{\mathcal{J}}^l$ denotes the set of left turn arcs, $\mathcal{A}_{\mathcal{J}}^{sf}$ denotes the set of straight forward arcs and $\mathcal{A}_{\mathcal{J}}^r$ is the set of all right turn arcs
- $\mathcal{A}_C$ denotes the set of arcs connecting intersections with sources and sinks
- $\mathcal{A}_S(co)$ is the set of arcs pointing from super source S to sources of commodity $co \in C$
- $\mathcal{A}_D(co)$ is set of arcs emerging from sinks of commodity $co \in C$ to the super sink D
- $\mathcal{A} = \mathcal{A}_{\mathcal{J}} \cup \mathcal{A}_C \cup \bigcup_{co \in C} \mathcal{A}_S(co) \cup \bigcup_{co \in C} \mathcal{A}_D(co)$ is the set of all arcs of network $\mathbf{G}$

In Fig. 7.3 a street network $\mathbf{G} = (\mathcal{N}, \mathcal{A})$ is depicted. Let node i_4 be an assembly point, indicated by an $\mathbf{A}$. Node i_4 is a source of commodities bus_{out} and veh and a sink of commodity bus_{in}. The nodes i_6 and i_7 denote nodes that represent safe zones. They are sources of bus_{out} and sinks of commodities bus_{in} and veh. The nodes $i_1, i_2, \ldots, i_5$ are sources of the commodity veh. Nodes $j_1, j_2, \ldots, j_8$ represent entrances/exits of intersections.

7.2.2 Parameters and Variables

Parameters and variables change only slightly. They are repeated shortly and the new parameters and variables are introduced:

- $T = \{0, 1, \ldots, T\}$ is the set of considered points in time with T is the upper bound on the number of time units
- $\tau(co, i, j)$ denotes the travel time from node i to j on arc $(i, j) \in \mathcal{A}_{\mathcal{J}} \cup \mathcal{A}_C \cup \mathcal{A}_S(co) \cup \mathcal{A}_D(co)$ of commodity $co \in C$
- $L(i, j)$ denotes the number of available lanes on arc $(i, j) \in \mathcal{A}_{\mathcal{J}} \cup \mathcal{A}_C$
- $O(co, s)$ stands for the number of commodity $co \in C$ in source $s \in \mathcal{S}(co)$ at the beginning of the time horizon, i.e. at point in time $t = 0$

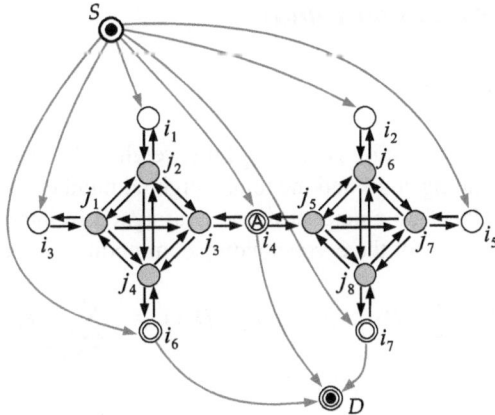

Fig. 7.3 Illustration of a street network **G** for the multicommodity model

- $Bout(s)$ is the number of buses bus_{out} that have to leave assembly point $s \in \mathcal{S}(bus_{out})$ to rescue the people waiting there
- $c^{in}(i,j)$ and $c(i,j)$ denote the inflow and the total capacity of arc $(i,j) \in \mathcal{A}_{\mathcal{J}} \cup \mathcal{A}_{\mathcal{C}}$ per lane and per time unit, respectively
- $\varsigma^{in}(co)/\varsigma^{in}_{\mathcal{J}}(co)$ denotes the consumption of the inflow capacity per unit of inflow of commodity $co \in \mathcal{C}$ outside/within intersections per time period
- $\varsigma(co)/\varsigma_{\mathcal{J}}(co)$ denotes the consumption of the total street capacity per time unit of one unit of flow of commodity $co \in \mathcal{C}$ on arcs outside/within intersections
- $\mathcal{BS}(s) \subset \mathcal{T}$ is the possible bus schedule for commodity $co \in \mathcal{C}_e$ in source $s \in \mathcal{S}(co)$ (it determines possible starting points in time)

The decision variables of the multicommodity model are the following:

- $x(co,i,j,t) \geq 0$ denotes the inflow of commodity $co \in \mathcal{C}$ on arc $(i,j) \in \mathcal{A}$ that leaves node i at point in time $t \in \mathcal{T}$ and enters node j at point in time $t + \tau(co,i,j)$ with $x(co,i,j,t) = 0$ for all $co \in \mathcal{C}, (i,j) \in \mathcal{A}, t \in \{-\tau(i,j), \ldots, -1\}$, for the ease of notation
- $\ell_e(i,j) \in \mathbb{N}_0$ represents the number of emergency-lanes assigned to arc $(i,j) \in \mathcal{A}_{\mathcal{J}}$ with $\ell_e(i,j) = 0$ for all $(i,j) \in \mathcal{A}^r_{\mathcal{J}}$
- $\ell_p(i,j) \in \mathbb{N}_0$ represents the number of public-lanes assigned to arc $(i,j) \in \mathcal{A}_{\mathcal{J}} \cup \mathcal{A}_{\mathcal{C}}$
- $y_g(i,j) \in \{0,1\}$ are binary variables with $y_g(i,j) = 1$ if it is allowed to assign a positive number of public-lanes (for $g = p$) and emergency-lanes (for $g = e$) to direction $(i,j) \in \mathcal{A}^l_{\mathcal{J}} \cup \mathcal{A}^{sf}_{\mathcal{J}}$ or if not ($= 0$)
- $n_e(q) \geq 0$ denotes the number of emergency-lanes within intersection $q \in \mathcal{J}$ with $\alpha_q \geq 2$ (we set $n_e(q) = 0$ for intersections q with $\alpha_q = 2$)

Right turn arcs do not cause crossing conflicts and emergency-vehicles (the buses) can use public-lanes too, therefore the variables $\ell_e(i,j)$ are set to zero for $(i,j) \in \mathcal{A}^r_{\mathcal{J}}$.

7.2.3 Objective and Constraints

Objective Function

The objective function minimizes the sum of the weighted flows of the commodities arriving in corresponding sinks and the total weighted number of emergency-lanes, where P_i is a penalty if an emergency-lane is assigned in intersection $i \in \mathcal{J}$ and Pr_{co} is a priority weight for the corresponding commodity $co \in \mathcal{C}$:

$$\min \sum_{co \in \mathcal{C}} \sum_{d \in \mathcal{D}(co)} \sum_{t \in \mathcal{T}} Pr_{co} \cdot t \cdot x(co, d, D, t) + \sum_{i \in \mathcal{J} : \alpha_i \geq 2} P_i \cdot n_e(i) \tag{7.1}$$

Network Flow Constraints

- Flow conservation constraints for the commodities bus_{in}, bus_{out}, veh:

$$\sum_{j \in \mathcal{N} : (j,i) \in \mathcal{A}} x(co, j, i, t - \tau(co, j, i)) = \sum_{j \in \mathcal{N} : (i,j) \in \mathcal{A}} x(co, i, j, t)$$

$$\text{for all } co \in \mathcal{C}; \ i \in \mathcal{N} \backslash \{S, D\}; \ t \in \mathcal{T} \tag{7.2}$$

- Avoidance of irregular behavior I: (within intersections)

$$\sum_{k \in \mathcal{N}_{\mathcal{J}} : (k,i) \in \mathcal{A}_{\mathcal{J}}} x(co, k, i, t - \tau(k, i)) = x(co, i, j, t)$$

$$\text{for all } co \in \mathcal{C}; \ i \in \mathcal{N}_{\mathcal{J}}; \ (i, j) \in \mathcal{A}_{\mathcal{C}}; \ t \in \mathcal{T} \tag{7.3}$$

- Avoidance of irregular behavior II:

$$x(co, i, j, t - \tau(i, j)) \leq \sum_{(j,k) \in (\mathcal{A}_{\mathcal{C}} \cup \mathcal{A}_{\mathcal{J}} \cup \mathcal{A}_D(co)) \backslash \{(j,i)\}} x(co, j, k, t)$$

$$\text{for all } co \in \mathcal{C}; \ (i, j) \in \mathcal{A}_{\mathcal{C}}; \ t \in \mathcal{T} \tag{7.4}$$

Vehicles of Evacuees

Constraints concerning the vehicles are the following:

- All vehicles of evacuees have to leave the corresponding source within the time horizon:

$$\sum_{t \in \mathcal{T}} x(veh, S, s, t) = O(veh, s) \quad \text{for all } s \in \mathcal{S}(veh) \tag{7.5}$$

- Every vehicle of an evacuee has to enter a safe zone within the time horizon:

$$\sum_{d \in \mathcal{D}(veh)} \sum_{t \in \mathcal{T}} x(veh, d, D, t) = \sum_{s \in \mathcal{S}(veh)} O(veh, s) \qquad (7.6)$$

Buses

Constraints concerning the buses are the following:

- A possible schedule of buses bus_{in} and bus_{out} that start at sources $\mathcal{S}(bus_{in})$ and $\mathcal{S}(bus_{out})$, respectively, is given:

$$x(co, S, s, t) \leq 1 \quad \text{for all } co \in \mathcal{C}_e; \ s \in \mathcal{S}(co); \ t \in \mathcal{BS}(s) \qquad (7.7)$$

$$x(co, S, s, t) = 0 \quad \text{for all } co \in \mathcal{C}_e; \ s \in \mathcal{S}(co); \ t \in \mathcal{T} \backslash \mathcal{BS}(s) \qquad (7.8)$$

- The number of buses that have to leave the assembly point $s \in \mathcal{S}(bus_{out})$ is $Bout(s)$ (to transport all pedestrians to a safe area):

$$\sum_{t \in \mathcal{T}} x(bus_{out}, S, s, t) = Bout(s) \quad \text{for all } s \in \mathcal{S}(bus_{out}) \qquad (7.9)$$

- All buses that are needed to transport the pedestrians to safe zones have to leave the evacuation zone within the time horizon:

$$\sum_{d \in \mathcal{D}(bus_{out})} \sum_{t \in \mathcal{T}} x(bus_{out}, d, D, t - \tau(bus_{out}, d, D)) = \sum_{s \in \mathcal{S}(bus_{out})} Bout(s)$$

$$(7.10)$$

The travel time $\tau(bus_{out}, d, D)$ could model the time needed to empty the bus and/or drive to a shelter and back to the evacuation zone. (The summand "$-\tau(bus_{out}, d, D)$" is only needed if $\tau(bus_{out}, d, D) > 0$ is valid for the artificial arc (d, D).)

- Capacity of number of buses bus_{out} at assembly point $a \in \mathcal{S}(bus_{out})$: The capacity of number of buses bus_{out} in $a \in \mathcal{S}(bus_{out})$ at point in time t depends on the number of buses that are assigned to a at point in time 0 (that is the value $O(bus_{out}, a)$), the number of buses that have left a until the point in time $t - 1$ and the number of buses that have entered a until the point in time $t - 1$.

$$x(bus_{out}, S, a, t) \leq O(bus_{out}, a) + \sum_{\tilde{t}=0}^{t-\tau(bus_{in}, a, D)-1} x(bus_{in}, a, D, \tilde{t})$$

$$- \sum_{\tilde{t}=0}^{t-1} x(bus_{out}, S, a, \tilde{t}) \quad \text{for all } a \in \mathcal{S}(bus_{out}); \ t \in \mathcal{T}$$

$$(7.11)$$

An assumption is that $\mathcal{D}(bus_{in}) = \mathcal{S}(bus_{out})$, but if not, i.e. if $a \notin \mathcal{D}(bus_{in})$, then the following constraint would be used (no other bus would enter the assembly place):

$$x(bus_{out}, S, a, t) \leq O(bus_{out}, a) - \sum_{\tilde{t}=0}^{t-1} x(bus_{out}, S, a, \tilde{t})$$

$$\text{for all } a \in \mathcal{S}(bus_{out}); \ t \in \mathcal{T}$$

- Capacity of number of buses bus_{in} at safe area $a \in \mathcal{S}(bus_{in})$:

$$x(bus_{in}, S, a, t) \leq O(bus_{in}, a) + \sum_{\tilde{t}=0}^{t-\tau(bus_{out}, a, D)-1} x(bus_{out}, a, D, \tilde{t})$$

$$- \sum_{\tilde{t}=0}^{t-1} x(bus_{in}, S, a, \tilde{t}) \quad \text{for all } a \in \mathcal{S}(bus_{in}); \ t \in \mathcal{T}$$

$$(7.12)$$

If $a \notin \mathcal{D}(bus_{out})$ then the following constraint would be used:

$$x(bus_{in}, S, a, t) \leq O(bus_{in}, a) - \sum_{\tilde{t}=0}^{t-1} x(bus_{in}, S, a, \tilde{t})$$

$$\text{for all } a \in \mathcal{S}(bus_{in}); \ t \in \mathcal{T}$$

The total number of buses is restricted with the constraints "Capacity of number of buses bus_{out}" (7.11) and "Capacity of number of buses bus_{in}" (7.12).

Traffic Routing Constraints

We differentiate between emergency- and public-lanes within intersections. For arcs in $\mathcal{A}_C$ only public-lanes are considered.

- The number of assigned lanes must not exceed the number of available lanes:

$$\ell_p(i, j) + \ell_p(j, i) \leq \mathcal{L}(i, j) \ \text{ for all } (i, j) \in \mathcal{A}_C : j \in \mathcal{N}_{\mathcal{J}}$$

$$(7.13)$$

$$\ell_p(i, j) + \ell_p(j, i) + \ell_e(i, j) + \ell_e(j, i) \leq \mathcal{L}(i, j) \ \text{ for all } (i, j) \in \mathcal{A}_{\mathcal{J}} \quad (7.14)$$

- Coupling constraints:

$$\ell_g(i, j) \leq \mathcal{L}(i, j) y_g(i, j) \ \text{ for all } (i, j) \in \mathcal{A}_{\mathcal{J}}^l \cup \mathcal{A}_{\mathcal{J}}^{sf}; \ g \in \{p, e\} \quad (7.15)$$

- Prohibition of crossing conflicts:
  Crossing conflicts within intersections are prohibited in the case of public-lanes.

  $$y_p(i, j) + y_p(k, h) \leq 1$$

  for all $(i, j), (k, h) \in \mathcal{A}^l_{\mathcal{J}} \cup \mathcal{A}^{sf}_{\mathcal{J}}$ that build a crossing conflict   (7.16)

  The crossing conflicts are understood as in the previous chapters and the constraints are built in the same way.

- Counting emergency-lanes:
  The number of emergency-lanes is counted per intersection. Crossing conflicts of an emergency-lane with public-lanes and with other emergency-lane are permitted but the emergency-lanes are counted and lead to a penalty in the objective function.

  $$\sum_{(i,j) \in \mathcal{A}_{\mathcal{J}}(q)} y_e(i, j) \leq n_e(q) \text{ for all } q \in \mathcal{J} : \alpha_q \geq 2 \qquad (7.17)$$

- The merging of lanes is restricted:

  $$\sum_{(k,i) \in A^*_{\delta,i}} (\ell_e(k, i) + \ell_p(k, i)) \leq \ell_p(i, j) + (\delta - 1)$$

  for all $q \in \mathcal{J}; \ i \in \mathcal{N}_{\mathcal{J}}(q); \ (i, j) \in \mathcal{A}_C; \ \delta \in \{1, \ldots, \alpha_j - 1\};$ (7.18)
  $A^*_{\delta,i} \in \{A \subseteq \mathcal{A}_{\mathcal{J}}(q) | \ |A| = \delta, A \ni (k, i) \text{ for a } k \in \mathcal{N}_{\mathcal{J}}(q)\}$

  where $\mathcal{N}_{\mathcal{J}}(q)$ is the set of nodes of intersection $q \in \mathcal{J}$, $\mathcal{A}_{\mathcal{J}}(q)$ is the set of arcs within intersection $q \in \mathcal{J}$ and $|A|$ denotes the cardinality of set A (number of elements of set A).

Inflow and Total Capacity Constraints

$\varsigma(co)/\varsigma_{\mathcal{J}}(co)$ and $\varsigma^{in}(co)/\varsigma^{in}_{\mathcal{J}}(co)$ denote the consumption of total capacity and inflow capacity on arcs between/within intersections of one unit of flow of commodity $co \in \mathcal{C} = \{bus_{in}, bus_{out}, veh\}$ per time unit, respectively.

- Inflow capacity constraints:
  The inflow is restricted within intersection (i.e. on arcs $\mathcal{A}_{\mathcal{J}}$) for buses and vehicles.

  $$\sum_{co \in \mathcal{C}} \varsigma^{in}_{\mathcal{J}}(co) x(co, i, j, t) \leq c^{in}(i, j)(\ell_e(i, j) + \ell_p(i, j))$$

  for all $(i, j) \in \mathcal{A}_{\mathcal{J}}; \ t \in \mathcal{T}$ (7.19)

Vehicles are only allowed to use public-lanes, hence the inflow capacity has to be restricted for vehicles depending on the number of lanes ℓ_p.

$$\varsigma_{\mathcal{J}}^{in}(veh)x(veh,i,j,t) \leq c^{in}(i,j)\ell_p(i,j) \quad \text{for all } (i,j) \in \mathcal{A}_{\mathcal{J}}; \, t \in \mathcal{T} \quad (7.20)$$

On connecting arcs $\mathcal{A}_C$ buses and vehicles can share all lanes and, hence, the inflow capacity constraints are the following:

$$\sum_{co \in \mathcal{C}} \varsigma^{in}(co)x(co,i,j,t) \leq c^{in}(i,j)\ell_p(i,j) \quad \text{for all } (i,j) \in \mathcal{A}_C; \, t \in \mathcal{T}$$

$$(7.21)$$

- Total street capacity constraints:
  The total street capacity is restricted on intersection arcs $\mathcal{A}_{\mathcal{J}}$ for buses and vehicles.

$$\sum_{co \in \mathcal{C}} \sum_{\tilde{t}=t-\tau(co,i,j)+1}^{t} \varsigma_{\mathcal{J}}(co)x(co,i,j,\tilde{t}) \leq c(i,j)(\ell_e(i,j) + \ell_p(i,j))$$

$$\text{for all } (i,j) \in \mathcal{A}_{\mathcal{J}}; \, t \in \mathcal{T} \quad (7.22)$$

Vehicles of evacuees are restricted to use only public-lanes within intersections.

$$\sum_{\tilde{t}=t-\tau(co,i,j)+1}^{t} \varsigma_{\mathcal{J}}(veh)x(veh,i,j,\tilde{t}) \leq c(i,j)\ell_p(i,j)$$

$$\text{for all } (i,j) \in \mathcal{A}_{\mathcal{J}}; \, t \in \mathcal{T} \quad (7.23)$$

The total street capacity associated with connecting arcs $\mathcal{A}_C$ is permitted to be shared by buses and vehicles.

$$\sum_{co \in \mathcal{C}} \sum_{\tilde{t}=t-\tau(co,i,j)+1}^{t} \varsigma(co)x(co,i,j,\tilde{t}) \leq c(i,j)\ell_p(i,j)$$

$$\text{for all } (i,j) \in \mathcal{A}_C; \, t \in \mathcal{T} \quad (7.24)$$

For every commodity different travel times may be chosen in the multicommodity urban evacuation model. That leads to the following discussion concerning the first-in-first-out property:

Positively speaking, we could say that the possibility of overtaking is provided. Especially, if we choose shorter travel times for buses or, if considering, for ambulances then commodities with a higher priority can always overtake commodities considered as a commodity with a lesser priority. I.e. besides the priority coefficient in the objective function, a priority can also be mapped with the travel times.

Negatively speaking, we could say that the first-in-first-out property is not captured properly. A commodity with a shorter travel time that enters a street section later than a commodity with a higher travel time may leave the street section before. Even on a narrow road where no overtaking is possible the faster commodity could leave the street before a slower commodity, even if it has entered the street section after the slower commodity. In our examples, we will use the same travel time for every commodity.

Note that if a crossing conflict occurs, it is possible that a vehicle and a bus travel through these crossing arcs in the same time. This case is not prohibited with this model.

7.3   An Illustration of Solutions of the Multicommodity Urban Evacuation Model

A trade off between both components of the objective function have to be decided, i.e. the values of the priority weights for each commodity and the penalty costs for the assignment of an emergency-lane. The first component is the weighted sum of flows entering the corresponding sinks, the second the weighted number of emergency-lanes. Using high penalty costs (compared to the priority weights of the commodities) may lead indirectly to longer routes to the corresponding sinks, using low penalty costs (compared to the priority weights of the commodities) may lead to an unclear traffic routing within intersections in terms of a larger number of emergency-lanes. Intersections known as confusing or dangerous can be set a high penalty for assigning an emergency-lane. If an emergency-lane is assigned and positive costs are set then at least one crossing conflict occurs. The assignment of no emergency-lane does not induce that no bus can travel through the intersection because buses can use every type of lane, i.e. public- and emergency-lanes.

We will give an example that is based on a 2×3 grid network and vary the penalty costs and priority weights. The model was implemented with AMPL using CPLEX 10.0 on a computer running Windows 7. The computation times of the considered instances with penalty costs $P_i = 1$ for intersections $i \in \mathcal{J}$ ($\alpha_i = 4$) were around 2 min, and the computation of the multicommodity model took around 1.4 h for instances with $P_i = 300$ for $i \subset \mathcal{J}$ ($\alpha_i = 1$).

The 2×3 grid network is depicted in Fig. 7.4. The values within the sources s denote the number of vehicles $O(veh, s)$ that have to leave the evacuation zone starting from s. The gray sources represent sources that are assigned to be assembly points where one bus has to leave each of them ($Bout(a) = 1$ for $a = a_1, a_2$). No bus is located at the assembly points in the beginning ($O(bus_{out}, a) = 0$ for $a = a_1, a_2$). The values next to the assembly nodes denote these values, i.e. $(Bout(a), O(bus_{out}, a)) = (1, 0)$ for both gray nodes $a \in \mathcal{S}(b_{out})$. The assembly nodes a_1 and a_2 are starting points for vehicles veh and buses bus_{out} that carry evacuees to a safe zone i_1, i_2 or i_3. These nodes are sinks for buses bus_{in} that travel from a safe zone to one of these two assembly nodes. One bus is placed outside

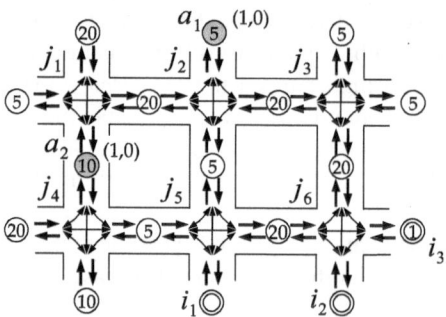

Fig. 7.4 Illustration of a street network based on a 2 × 3 grid network

the evacuation zone, in the safe zone i_3 on the left. The safe zones are indicated with a double lined node. Each of the safe zone nodes may be starting points for the commodity bus bus_{in} and sinks for the commodities vehicle veh and buses bus_{out}. The example is constructed such that the bus located in the safe zone has to enter the evacuation area twice. The parameters were chosen as follows:

- The travel time of all commodities $co \in C$ is $\tau(co, i, j) = 5$ time units for arcs $(i, j) \in A_C$ and $\tau(co, i, j) = 1$ time units for arcs $(i, j) \in A_J$.
- The inflow capacity for all arcs $A_C \cup A_J$ is $c^{in}(i, j) = 1$ per lane and per time unit.
- The total capacities are $c(i, j) = 75$ for all arcs $(i, j) \in A_C$ and $c(i, j) = 15$ for all $(i, j) \in A_J$ per lane and per time unit.
- The consumption coefficients $\varsigma^{in}(co)$ and $\varsigma^{in}_J(co)$ are set to one for all $co \in C$ per lane. The coefficients $\varsigma(co)/\varsigma_J(co)$ are set to 18/7 per lane for vehicles and 26/15 per lane for buses.
- The available number of lanes $\mathcal{L}(i, j)$ is set to three for every arc $(i, j) \in A_C \cup A_J$.

Example: $P_i = 1$ for all $i \in J$ ($\alpha_i = 4$) and $Pr_{co} = 1$ for all $co \in C$

The penalty costs are assumed to be the same for every intersection. First, an optimal solution of the instance with penalty cost $P_i = 1$, $i \in J$ with $\alpha_i = 4$ and with the priority value $Pr_{co} = 1$ for all $co \in C$ is considered. All commodities have the same priority. The optimal value of the objective function is 5035.67 where four units are the weighted sum of number of emergency lanes: here with penalty cost of one, it is equal to number of emergency-lanes. Figure 7.5a depicts the number of lanes of an optimal solution, each arc represents a lane. The black arcs depict emergency-lanes, to be seen in intersections j_1, j_2 (here are two emergency-lanes assigned) and j_6. The last vehicles leave the evacuation zone in time unit 46 through exit i_1, the bus leaves the evacuation zone the second and last time in time unit 137. The bus enters the evacuation zone trough the entrance i_3, travels through intersections j_6, j_5 and j_2 to enter the assembly point a_1 to pick up the evacuees and travel back

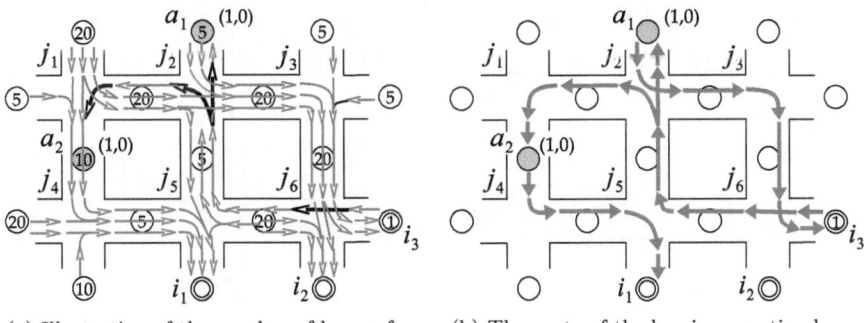

(a) Illustration of the number of lanes of an optimal solution

(b) The route of the bus in an optimal solution

Fig. 7.5 Illustration of an optimal solution with penalty cost $P_i = 1$ for all i and priority costs $Pr_{co} = 1$ for all co

through intersection j_2, j_3, j_6 to safe zone i_3. The second time, the bus starts in i_3, traverses intersections j_6, j_5, j_2, j_1, to enter assembly point a_2 and travels to through intersections j_4, j_5 to safe zone i_1. The route of the bus is depicted in Fig. 7.5b.

Example: $P_i = 300$ for all $i \in \mathcal{J}$ $(\alpha_i = 4)$ and $Pr_{co} = 1$ for all $co \in \mathcal{C}$

In the second instance based on the above introduced example, the penalty costs are varied. An optimal solution of the instance with penalty cost $P_i = 300$ for all $i \in \mathcal{J}$ $(\alpha_i = 4)$ for assigning an emergency-lane and with the priority cost $Pr_{co} = 1$ for all commodities $co \in \mathcal{C}$ is considered. The value of the objective function is $5,379.83$, where no emergency-lanes are assigned, i.e. no crossing conflict occurs in this solution. Figure 7.6a depicts the number of lanes of an optimal solution. The last vehicles leave the evacuation zone through exits i_1, i_2 and i_3 in time units 50, 49 and 39, respectively. The bus leaves the evacuation zone the second and last time in time unit 152 (i.e. the bus leaves the evacuation zone later than in the first example). The bus route is illustrated in Fig. 7.6b. The bus enters the evacuation zone trough the entrance i_3, travels through intersections j_6, j_3 and j_2 to enter the assembly point a_1 to pick up the evacuees and travel back through intersections j_2, j_5, j_6 to safe zone i_3. The second time, the bus starts in i_3, uses intersections j_6, j_3, j_2, j_5, j_4, enters assembly point a_2 and travels through intersections j_4, j_5 back to safe zone i_1. Note that between the intersections j_2 and j_5, the bus has to order in the appropriate lane such it can turn without causing a crossing conflict.

Increasing the priority costs for buses to 10 then the solution of the above examples changes such that the evacuation time decreases, i.e. the bus leaves the evacuation zone the second time earlier than in instances before. The results are illustrated in the following example.

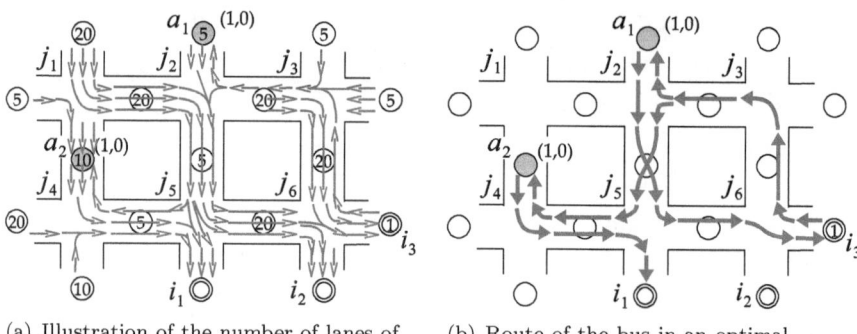

(a) Illustration of the number of lanes of   (b) Route of the bus in an optimal
an optimal solution   solution

Fig. 7.6 Illustration of an optimal solution with penalty cost $P_i = 300$ for all i and priority costs $Pr_{co} = 1$ for all co

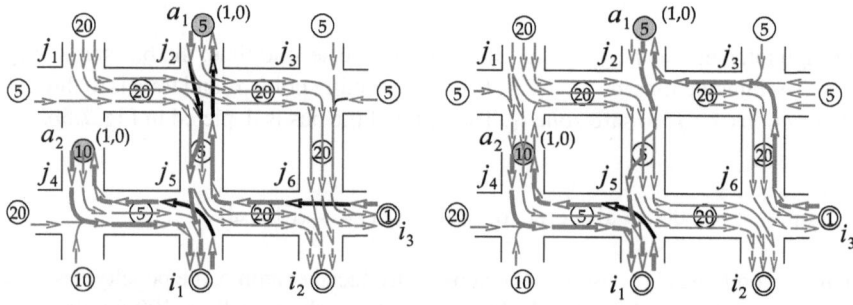

(a) Number of lanes of an optimal solu-   (b) Number of lanes of an optimal solution with penalty cost $P_i = 1$ for all i tion with penalty cost $P_i = 300$ for all i

Fig. 7.7 Illustration of an optimal solution with different penalty costs and priority costs $Pr_{co} = 10$ for all $co \in C_e$, $Pr_{veh} = 1$

Examples: $P_i = 1/P_i = 300$ for all $i \in \mathcal{J}$ $(\alpha_i = 4)$ and $Pr_{veh} = 1$, $Pr_{co} = 10$ for all $co \in C_e$

Let the costs be $P_i = 1$ for $i \in \mathcal{J}$ with $\alpha_i = 4$. The chosen priority costs are $Pr_{co} = 10$ for the emergency vehicles $co \in C_e$ and $Pr_{veh} = 1$ for the vehicles of the evacuees. The optimal objective function value is $7,742.33$ with the penalty four (four emergency-lanes are assigned). The value of the weighted sum of flows with costs $Pr_{co} = 1$ for all commodities of this solution is $\sum_{co \in C} \sum_{t \in T} \sum_{(d,D) \in \mathcal{A}_D(co)} t \cdot x(co, d, D, t) = 5,227.33$ (the value can be compared to the weighted sum of flows achieved in the first example, it is $5,031.67$). The number of lanes of the optimal solution are illustrated in Fig. 7.7a. The black arcs denote the emergency-lanes that are assigned in intersections j_2 twice and in intersections j_5 and j_6. The last car arrives in a safe zone in time unit 50, the bus arrives from the second trip in time period 107. The route of the bus is depicted in Fig. 7.7a as bold arcs. The route of bus_{in} starts in i_3 using intersections j_6, j_5 and j_2 to enter sink a_1. As bus_{out} the bus travels from a_1 through intersection j_2, orders in the appropriate lane on the way to j_5 to transport the evacuees to sink i_1. There the bus starts again as bus_{in} using intersections j_5 and j_4 to load evacuees in assembly node a_2 and travels back as bus_{out} to the sink (of the bus_{out} commodity) i_1.

Changing the penalty costs to $P_i = 300$, $i \in \mathcal{J}$ ($\alpha_i = 4$), then the evacuation time does not change, it stays 107 with the bus entering the second and last time a safe zone. The value of the optimal objective function is 8, 324.33 with the penalty of 300 for one emergency-lane. The value of the weighted sum of flows of this solution is 5, 513.33 with costs $Pr_{co} = 1$ for all commodities $co \in \mathcal{C}$. The last car arrives in a safe zone in time unit 52, and one emergency-lane in intersection j_5 leads to one crossing conflict. The number of lanes of the optimal solution, the emergency-lane that is assigned in intersection j_5 and the route of the bus are depicted in Fig. 7.7b.

Chapter 8
A Four-Staged Heuristic Approach to Solve the Urban Multicommodity Model

The heuristic solution approach for the multicommodity model that is presented in Sect. 7.2 consists of four steps. The first two steps compute flows and a traffic routing on the reduced network, Step 3 on subgraphs for every single intersection and Step 4 works with the complete network including the intersection subgraphs.

The solution of the routes of the emergency vehicles is determined before incorporating the vehicles of the evacuees. These two steps are operated on the reduced network without the subgraphs of the intersections. The reduced network substitutes a single node for a subgraph of every intersection (compare Sect. 6.1). In Fig. 8.1 a reduced network and a detailed network of two intersections with $\alpha = 4$ entrances/exits and six intersections with $\alpha = 1$, a dead end or an entrance to a safe zone, are illustrated.

In Step 3 the results of the number of lanes of the arcs $\mathcal{A}_C$ and the flows of the commodities resulting from Step 2 are used to compute the number of lanes within every intersection individually. The flow within the intersections should be similar to the flow of Step 2 while the constraints that prohibit crossing conflicts and restrict merging conflicts are included in this step. Step 4 computes the flow of all commodities in the complete network. The objective function value and the evacuation time of the traffic routing computed with Steps 1, 2 and 3 are determined by solving the multicommodity evacuation problem with fixed values for the number of lanes.

An overview of the four-staged heuristic approach is given in the following:

Step 1: The flow of the emergency vehicles with corresponding constraints is considered and the number of lanes for the emergency flow for every street section is determined in the reduced network. For the emergency vehicles at most one lane per street section is assigned.

Step 2: In Step 2 the evacuation flow (the vehicles of the evacuees) is included. The flow of all commodities is computed under the corresponding constraints and the number of lanes for the evacuation vehicles has to be determined with the following additional restrictions:

- The numbers of lanes that were determined in Step 1 are lower bounds for the corresponding street sections.

S. Bretschneider, *Mathematical Models for Evacuation Planning in Urban Areas*,
Lecture Notes in Economics and Mathematical Systems 659,
DOI 10.1007/978-3-642-28759-6_8, © Springer-Verlag Berlin Heidelberg 2013

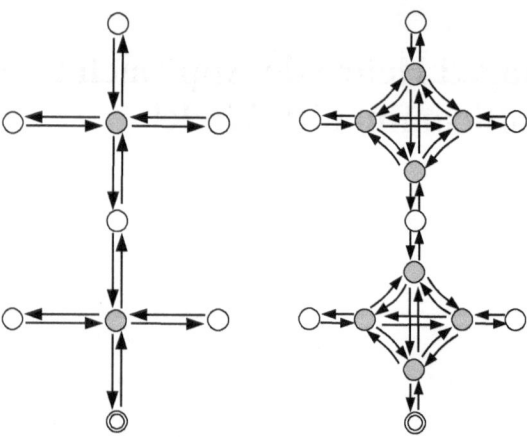

Fig. 8.1 Illustration of a reduced and detailed street network (without super source, super sink and the corresponding arcs)

- The emergency vehicles are allowed to use solely the routes computed in Step 1.

  Steps 1 and 2 are solved on the network without intersection arcs, i.e. crossing conflicts and restrictions of merging conflicts within intersections are not considered. The emergency inflow, the evacuation inflow and the number of lanes of the arcs $\mathcal{A}_C$ that connect intersections with sources and sinks are determined. After the computation of Step 1 and Step 2, the computation of the inflows of all commodities and the number of lanes within the intersections is undertaken. We distinguish between emergency-lanes and public-lanes. Certain parameters of Steps 3 and 4 depend on the results of Steps 1 and 2.

Step 3: Every intersection is considered separately: an intersection model incorporating the results of Step 2 is solved on every single subgraph associated with an intersection. The constraints of intersections, like the prohibition of crossing conflicts or the restriction of merging of lanes, are taken into account. The flows of the commodities should be similar to the foregoing results (i.e. the flows computed in Step 2).

Step 4: The number of lanes of connecting arcs $\mathcal{A}_C$ and intersection arcs $\mathcal{A}_J$ generated with Step 2 and Step 3 are fixed in the multicommodity evacuation model and the inflows of all commodities and the objective function value are computed.

The penalty for the assignment of emergency-lanes is taken into account in Step 3 considering every intersection separately, i.e. the optimization of the assignment of emergency-lanes is not considered onto the entire network.

In the following the four steps of the heuristic, the models of Steps 1, 2 and 3, in particular, will be presented. The heuristic approach will be illustrated with a small instance throughout this chapter.

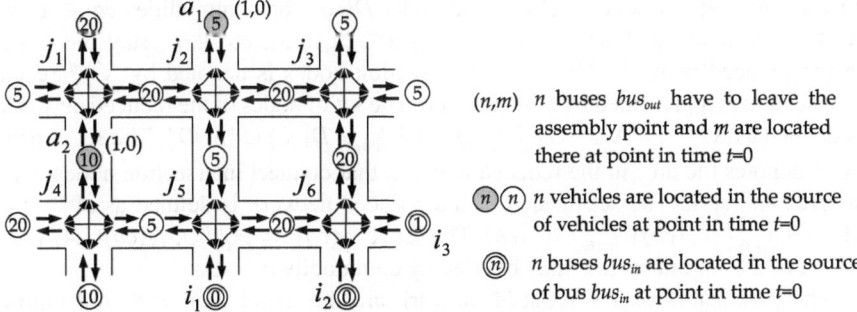

Fig. 8.2 Illustration of the network and number of evacuees of the 2 × 3-Example

The **2 × 3**-Example

The example that illustrates the four-staged heuristic approach is based on the 2 × 3 grid network which is introduced in Sect. 7.3. The same parameters are chosen for the inflow capacities c^{in}, the total street capacities c, the travel times τ, the consumption coefficients ς^{in} and $\varsigma_{\mathcal{J}}^{in}$ of the inflow capacity as well as the consumption coefficients ς and $\varsigma_{\mathcal{J}}$ of the total capacity. We assume penalty costs $P_i = 1$ for all $i \in \mathcal{J}$ with $\alpha_i = 4$ and a priority coefficient of $Pr_{co} = 1$ for all commodities $co \in \mathcal{C}$. See Fig. 8.2 for an illustration of the street network. Sources of vehicles are indicated by nodes drawn with a single line (white and gray), the safe zones are indicated with a double lined node and the assembly points as gray nodes.

8.1   First Stage of the Multicommodity Heuristic

The optimization model of Step 1 could include different properties the routes of the emergency vehicles should have. That could be the shortest routes from sources to sinks or the routes that cause the minimization of the average arrival time for every single emergency vehicle. The chosen model of Step 1 is directly generated from the provided multicommodity urban evacuation model (see Sect. 7.2). It minimizes the weighted sum of flows entering the corresponding sinks. With the following notations of parameters, sets and decision variables the model of Step 1 of the heuristic will be formulated.

Parameters and Sets

Every parameter and set that differs from the detailed network presentation of the multicommodity evacuation model is subscripted by red (reduced network). Let $\mathbf{G}^{red} = (\mathcal{N}^{red}, \mathcal{A}^{red})$ denote the reduced network. The super source S, the super sink D, the sets of sources $\mathcal{S}(co)$ and sinks $\mathcal{D}(co)$ for commodities $co \in \mathcal{C}$ as well as the associated set of arcs $\mathcal{A}_S(co)$ and $\mathcal{A}_D(co)$, $co \in \mathcal{C}$, stay the same in the reduced network. The set of intersection nodes is denoted by $\mathcal{N}_{\mathcal{J}}^{red}$, each intersection is represented by a single node. The set of nodes of the reduced network is defined as $\mathcal{N}^{red} := \mathcal{N}_{\mathcal{J}}^{red} \cup \bigcup_{co \in \mathcal{C}} \mathcal{S}(co) \cup_{co \in \mathcal{C}} \mathcal{D}(co) \cup \{S, D\}$. The set of arcs $\mathcal{A}_C^{red}$ denotes the arcs in the reduced network that connect intersection nodes with sources and sink. The set of arcs of the reduced network is defined as $\mathcal{A}^{red} := \mathcal{A}_C^{red} \cup \bigcup_{co \in \mathcal{C}} \mathcal{A}_S(co) \cup_{co \in \mathcal{C}} \mathcal{A}_D(co)$. The set $\mathcal{A}^{red}(co) := \mathcal{A}_C^{red} \cup \mathcal{A}_S(co) \cup \mathcal{A}_D(co)$ indicates the set of arcs that may be used by commodity $co \in \mathcal{C}$.

The parameters on the reduced network are the travel time τ^{red}, the inflow capacity c_{in}^{red}, the total capacity c^{red} and the usable number of lanes $\mathcal{L}^{red}$.

Decision Variables

The decision variables of the model of Step 1 are subscripted by $1st$. These are the real-valued non-negative inflow variables $x^{1st}(co, i, j, t) \in \mathbb{R}_0^+$ of the buses $co \in \mathcal{C}_e = \{bus_{in}, bus_{out}\}$ defined on arcs $(i, j) \in \mathcal{A}^{red}$ for $t \in \mathcal{T}$ (and let $x^{1st}(co, i, j, t) := 0$ for $t < 0$) and the binary decision variables $\ell^{1st}(i, j) \in \{0, 1\}$ denoting the number of lanes for $(i, j) \in \mathcal{A}_C^{red}$.

Objective Function

The objective is to minimize the weighted sum of flows of the emergency commodities $co \in \mathcal{C}_e$ that enter the arcs $(d, D) \in \mathcal{A}_D(co)$. The weights are increasing in time and consist of the priority value Pr_{co} for commodity $co \in \mathcal{C}_e$ and the value t of the associated point in time $t \in \mathcal{T}$.

$$\sum_{co \in \mathcal{C}_e} \sum_{t \in \mathcal{T}} \sum_{d \in \mathcal{D}(co)} Pr_{co} \cdot t \cdot x^{1st}(co, d, D, t) \qquad (8.1)$$

Network Flow Constraints

The network flow constraints of the model of Step 1 are analogously formulated as the network flow constraints of the multicommodity evacuation model. But the constraints for the first stage are only defined on the reduced network $\mathbf{G}^{red}$ and only for the emergency vehicles $co \in \mathcal{C}_e$ with the corresponding decision variables x^{1st} defined for Step 1.

- Flow conservation constraints (compare constraints (7.2)):

$$\sum_{(k,i)\in\mathcal{A}^{red}(co)} x^{1st}(co,k,i,t-\tau^{red}(co,k,i)) = \sum_{(i,k)\in\mathcal{A}^{red}(co)} x^{1st}(co,i,k,t)$$

(8.2)

$$\text{for all } co \in \mathcal{C}_e; \ i \in \mathcal{N}^{red}\backslash\{S,D\}; \ t \in \mathcal{T}$$

- Avoidance of irregular flow II (compare constraints (7.4)):

$$x^{1st}(co,i,j,t-\tau^{red}(co,i,j)) \leq \sum_{(j,k)\in\mathcal{A}_C^{red}\cup\mathcal{A}_D(co)\backslash\{(j,i)\}} x^{1st}(co,j,k,t) \quad (8.3)$$

$$\text{for all } co \in \mathcal{C}_e; \ (i,j) \in \mathcal{A}_C^{red}; \ t \in \mathcal{T}$$

Buses

The constraints concerning the commodity of buses of the multicommodty evacuation model (Sect. 7.2) are taken into account. They have to be adopted (like the constraints (8.2) and (8.3)) to the reduced network $\mathbf{G}^{red} = (\mathcal{N}^{red}, \mathcal{G}^{red})$ and the decision variables $x^{1st}(co,i,j,t)$ of the model of Step 1 with $co \in \mathcal{C}_e$, $(i,j) \in \mathcal{A}^{red}$, $t \in \mathcal{T}$. In contrast to the multicommodity evacuation model an additional constraint is needed to avoid unnecessary buses and unnecessary assigned lanes that lead to unintended effects in Step 2. This is constraint (8.4) that allows only an exact number of buses bus_{in} to enter the evacuation zone. Otherwise, more than necessary buses may cycle or travel in the network because additional buses do not lead, necessarily, to an increase of the value of the objective function. In Step 2 the arcs that are traversed by flow in the optimal solution of the model of Step 1 have to be identified but only these arcs that are traveled by buses needed to transport evacuees. To make the identification of these arcs easier in Step 2, constraint (8.4) is introduced in Step 1. The constraints that restrict the flow of the bus commodities are summarized in the following:

- A bus schedule that provides the points in time when the buses are allowed to leave their appropriate sources is given (analogously to constraints (7.7) and (7.8)).
- In total $Bout(s)$ buses bus_{out} have to leave the assembly point $s \in \mathcal{S}(bus_{out})$ to transport all pedestrians to a safe area (formulated analogously to (7.9)).
- All buses $\sum_{s\in\mathcal{S}(bus_{out})} Bout(s)$ that have to leave the assembly points $s \in \mathcal{S}(bus_{out})$ have to enter a safe zone within the time horizon (formulated analogously to (7.10)).

- The capacity of number of buses bus_{out} at assembly point $a \in S(bus_{out})$ has to be regarded, i.e. the number of buses bus_{out} that is ready to leave $a \in S(bus_{out})$ in point in time $t \in T$ is restricted (formulated analogously to (7.11)).
- The capacity of number of buses bus_{in} at safe area $a \in S(bus_{in})$ is taken into account, i.e. the number of buses bus_{in} that is ready to leave $a \in S(bus_{in})$ in point in time $t \in T$ is restricted (formulated analogously to (7.12)).
- The number of buses that have to enter from sources $S(bus_{in})$ the evacuation zone is equal to $\sum\limits_{s \in S(bus_{out})} (Bout(s) - O(bus_{out}, s))$:

$$\sum_{t \in T} \sum_{s \in S(bus_{in})} x^{1st}(bus_{in}, S, s, t) = \sum_{s \in S(bus_{out})} (Bout(s) - O(bus_{out}, s)) \quad (8.4)$$

Traffic Routing Constraints

The number of the available lanes $\mathcal{L}^{red}(i, j)$ associated with the arcs $(i, j) \in \mathcal{A}_C^{red}$ must not be exceeded by the assigned number of lanes $\ell^{1st}(i, j) + \ell^{1st}(j, i)$, $(i, j) \in \mathcal{A}_C^{red}$ (see constraint (7.13)).

Inflow and Total Capacity Constraints

The inflow and total street capacity constraints (7.21) and (7.24) concerning the arcs in $\mathcal{A}_C$ that connect intersections with sources and sinks are adopted to the reduced network, i.e. the constraints have to be formulated with arcs in $\mathcal{A}_C^{red}$ and the corresponding parameters ς_{in}, ς, c_{in}^{red} and c^{red} as well as the variables x^{1st} and ℓ^{1st}.

The 2×3-Example: Step 1

The instance of the 2×3-example of the Step 1 model is solved optimally. The value of the optimal function is 270. The numbers of lanes of the optimal solution of the example are depicted in Fig. 8.3, where each arc represents a lane. The bus needs 105 time units to complete both trips to the two assembly points a_1 and a_2 and arrive back at the safe zone i_1. The bus starts in safe zone i_3, traverses through intersections j_6, j_5 and j_2 to assembly point a_1 and carries the evacuees using intersections j_2 and j_5 to safe zone i_1. The second trip of the bus begins in safe zone i_1. It traverses through intersections j_5 and j_4 to assembly point a_2 and back to safe zone i_1 using intersections j_4 and j_5.

The assigned lanes build parameters for the model in Step 2: The numbers of lanes determined with Step 1 are lower bounds for the number of lanes of Step 2.

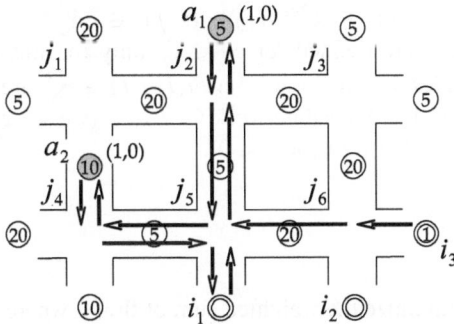

Fig. 8.3 Illustration of a solution of Step 1 of the 2 × 3-Example

Additionally, the commodity of the buses bus_{in} and bus_{out} in Step 2 are only defined on the arcs that they traveled in the solution of the model of Step 1.

8.2 Second Stage of the Multicommodity Heuristic

In Step 2 the flow of vehicles is incorporated. The inflow of vehicles is defined on the entire reduced network, whilst the inflow of the bus commodities is only defined on the arcs that are traversed by them in the solution of the model of Step 1.

The model of Step 2 is similar to the multicommodity evacuation model. It is only considered on the reduced network and additional the flow of buses $co \in C_e$ is restricted on arcs $\mathcal{A}^{red}_{used}(co)$ resulting from Step 1.

Parameters and Sets

The parameters of the model of Step 2 are defined on the reduced network $G^{red} = (\mathcal{N}^{red}, \mathcal{A}^{red})$, too. After solving the model of Step 1, the arcs that are traveled by commodity $co \in C_e$ in the solution of the model of Step 1 are identified and united into a set. Let $\mathcal{A}^{red}_{used}(co)$ denote this set of arcs that are used by commodity $co \in C_e$ in the solution achieved in Step 1. I.e. $\mathcal{A}^{red}_{used}(co)$, $co \in C_e$, includes the arcs $(i, j) \in \mathcal{A}^{red}(co)$ with $\sum_{t \in T} x^{1st}(co, i, j, t) > 0$. The inflow of the vehicles is defined on the set $\mathcal{A}^{red}(veh) := \mathcal{A}^{red}_C \cup \mathcal{A}_S(veh) \cup \mathcal{A}_D(veh)$.

Decision Variables

The variables are the number of lanes on the reduced street network also considered in Step 2, i.e. $\ell^{2nd}(i, j)$ for $(i, j) \in \mathcal{A}^{red}_C$. An additional lower bound is introduced for the number of lanes that results from the solution of the number of lanes of Step 1: $\ell^{1st}(i,j) \leq \ell^{2nd}(i,j)$, $(i,j) \in \mathcal{A}_C^{red}$. The inflow variables $x^{2nd}(co,i,j,t) \in \mathbb{R}_0^+$ are defined for $co \in \mathcal{C}_e$ only for arcs $(i,j) \in \mathcal{A}_{used}^{red}(co)$ and the inflow variables of vehicles $x^{2nd}(veh,i,j,t) \in \mathbb{R}_0^+$ are defined for all arcs $(i,j) \in \mathcal{A}^{red}(veh)$ of the reduced network for $t \in \mathcal{T}$ (we set $x^{2nd}(co,i,j,t) := 0$ for all $co \in \mathcal{C}$, $(i,j) \in \mathcal{A}^{red}(co)$ and $t < 0$).

Objective Function

The objective is to minimize the weighted sum of flows, where Pr_{co} is the priority weight for commodity $co \in \mathcal{C}$:

$$\sum_{t \in \mathcal{T}} \sum_{(d,D) \in \mathcal{A}_D(co)} Pr_{veh} \cdot t \cdot x^{2nd}(veh,d,D,t) \tag{8.5}$$

$$+ \sum_{co \in \mathcal{C}_e} \sum_{t \in \mathcal{T}} \sum_{(d,D) \in \mathcal{A}_{used}^{red}(co)} Pr_{co} \cdot t \cdot x^{2nd}(co,d,D,t)$$

Flow Conservation Constraints

- The flow conservation constraints have to be valid for every commodity $co \in \mathcal{C}$ for every node $i \in \mathcal{N}^{red} \setminus \{S,D\}$ in every point in time $t \in \mathcal{T}$, where only the corresponding defined arcs for the commodities, i.e. $\mathcal{A}^{red}(veh)$, $\mathcal{A}_{used}^{red}(bus_{in})$ and $\mathcal{A}_{used}^{red}(bus_{out})$ for the commodities veh, bus_{in} and bus_{out}, respectively, are taken into account (analogously to (8.2)).
- The constraints of "Avoidance of irregular flow II" have to be valid for every commodity $co \in \mathcal{C}$ on the corresponding defined street network arcs in every point in time (analogously to (8.3)).

Vehicles

The flow enforcement constraints for vehicles are formulated for the reduced network.

- Every vehicle assigned to a source has to start within time horizon $\mathcal{T}$ (analogously to constraint (7.5)).
- All vehicles have to reach a sink within the given time horizon (analogously to constraint (7.6)).

Buses

The inflow of buses $co \in C_e$ is only defined on $\mathcal{A}_{used}^{red}(co)$, hence the following constraints only consider these arcs.

- A bus schedule is provided with possible starting points in time (see constraints 7.7) and (7.8)).
- $Bout(a)$ buses bus_{out} have to start from the assembly point $a \in \mathcal{S}(bus_{out})$ to carry all the pedestrians to safe zones (analogously to constraints (7.9)).
- The number of buses that have to reach the safe zones are the total number of all buses that have to leave the sources $\mathcal{S}(bus_{out})$, i.e. $\sum\limits_{a \in \mathcal{S}(bus_{out})} Bout(a)$ (analogously to constraints (7.10)).
- The capacity of the number of buses at the sources of the considered commodity are restricted on the number of buses that are assigned to the source at the beginning of the time horizon as well as the number of buses that have left and entered the node until the considered point in time (analogously to (7.11) and (7.12)).
- The number of buses that have to enter the evacuation zone is restricted. This constraint is not needed in the urban multicommodity model, but in the model of Step 2. It is needed to assure that there are not more buses than necessary within the network such that in Step 3 feasibility can be guaranteed (analogously to (8.4)).

$$\sum_{(S,a) \in \mathcal{A}_{used}^{red}(bus_{in})} \sum_{t \in T} x^{2nd}(bus_{in}, S, a, t) \qquad (8.6)$$

$$= \sum_{a \in \mathcal{S}(bus_{out})} ((Bout(a) - O(bus_{out}a))$$

Traffic Routing Constraints

- Using results from Step 1, the lanes traversed by flow in the considered solution of the model of Step 1 are a lower bounds for the number of lanes in Step 2. The number of lanes of arcs $(i, j) \in \mathcal{A}_C^{red}$ with $\ell^{1st}(i, j) = 1$ are considered: If $\sum_{co \in C_e} \sum_{t \in T} x^{1st}(co, i, j, t) = 0$ then we set $\ell^{1st}(i, j) := 0$, otherwise the value remains $\ell^{1st}(i, j) = 1$, as computed in Step 1. Then the lower bounds of the number of lanes in Step 2 are restricted as follows

$$\ell^{1st}(i, j) \le \ell^{2nd}(i, j) \text{ for all } (i, j) \in \mathcal{A}_C^{red} \qquad (8.7)$$

- The available number of lanes $\mathcal{L}^{red}$ must not be exceeded by the assigned number of lanes of arcs associated with a street segment, i.e. $\ell^{2nd}(i, j) + \ell^{2nd}(j, i) \le \mathcal{L}^{red}(i, j)$ for $(i, j) \in \mathcal{A}_C^{red}$ (compare constraints (7.13)).

Inflow and Total Capacity Constraints

For the inflow and total capacity constraints, different cases according to the set of arcs of the different commodities have to be considered. The inflow variable of commodity veh is defined on the entire street network denoted by $\mathcal{A}_C^{red}$ and the artificial super arcs emerging from and pointing to S and D, i.e. $\mathcal{A}_S(veh)$ and $\mathcal{A}_D(veh)$, respectively.

- The inflow capacity on arcs that are defined only for vehicles is restricted for every point in time as follows:

$$\varsigma^{in}(veh) \cdot x^{2nd}(veh, i, j, t) \leq c_{in}^{red}(i, j) \cdot \ell^{2nd}(i, j) \tag{8.8}$$

$$\text{for all } (i, j) \in \mathcal{A}_C^{red} \setminus \bigcup_{co \in C_e} \mathcal{A}_{used}^{red}(co); t \in T$$

The inflow capacity is restricted on arcs that are defined for all commodities $co \in C$ as follows:

$$\sum_{co \in C} \varsigma^{in}(co) \cdot x^{2nd}(co, i, j, t) \leq c_{in}^{red}(i, j) \cdot \ell^{2nd}(i, j) \tag{8.9}$$

$$\text{for all } (i, j) \in \bigcap_{co \in C_e} \mathcal{A}_{used}^{red}(co) \setminus \left(\bigcup_{co \in C_e} \mathcal{A}_S(co) \cup \mathcal{A}_D(co) \right); t \in T$$

For the last two cases, the inflow capacity per point in time has to be regarded on arcs that are defined for vehicles and exactly one other commodity:

$$\sum_{\tilde{co} \in \{co, veh\}} \varsigma^{in}(\tilde{co}) \cdot x^{2nd}(\tilde{co}, i, j, t) \leq c_{in}^{red}(i, j) \cdot \ell^{2nd}(i, j) \tag{8.10}$$

$$\text{for all } co \in C_e; t \in T;$$

$$(i, j) \in \mathcal{A}_{used}^{red}(co) \setminus \left(\bigcup_{\hat{co} \in C_e \setminus \{co\}} \mathcal{A}_{used}^{red}(\hat{co}) \cup \mathcal{A}_S(\hat{co}) \cup \mathcal{A}_D(\hat{co}) \right)$$

- The constraints for the total capacity on arcs are built analog. The total capacity has to be regarded on every arc of the street network for every point in time. The four cases, arcs that are only defined for the inflow of vehicles, arcs that can be used by all commodities and arcs that can only be used by vehicles and exactly one other commodity (there are two bus commodities bus_{in} and bus_{out}) have to be considered (see constraint (7.24)).

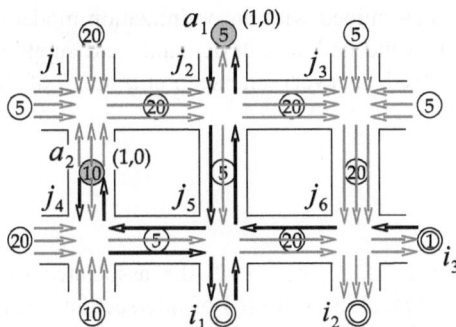

Fig. 8.4 Illustration of the number of lanes of the solution resulting from Step 2 of the 2 × 3-example

The results of the solution of the model of Step 2 that are used for Step 3 are the number of lanes of every connecting arc in $\mathcal{A}_C^{red}$ and the flow of the different commodities entering or leaving the intersections.

The 2 × 3-Example: Step 2

Based on the results of Step 1, the model of Step 2 is built. The number of lanes of arcs that are traversed by flow in the solution of Step 1 are restricted from below by one lane in Step 2. The optimal objective value of the Step 2 model of the 2 × 3-example is $5,031.33$. The resulting number of lanes are depicted in Fig. 8.4. Each arc denotes an assigned lane, the bold black arcs mark the lanes that result from Step 1.

The bus enters the safe zone finally in time unit 105. The last cars arrive at the safe zones i_1, i_2 and i_3 in time units 47, 46 and 46, respectively. The bus travels from safe zone i_3 using intersections j_6, j_5 and j_4 to assembly point a_2, carries the evacuees to safe zone i_1 (using intersections j_4 and j_5) and drives through i_5 and j_2 to assembly point a_1 and back to safe zone i_1.

The results that are used as parameters in Step 3 are the numbers of lanes of the arcs $\mathcal{A}_C^{red}$ connecting intersections with sources and sinks and the total flows of each commodity that travel through these arcs over the total time horizon. In Step 3, the numbers of lanes within the intersections are determined.

8.3   Third Stage of the Multicommodity Heuristic

In Step 3 every intersection is considered separately. With each intersection $q \in \mathcal{J}$, $\alpha_q \geq 2$, a graph is associated that depends on results of Step 2. Let q with $q \in \mathcal{J}$ denotes the intersection node in the reduced network. The numbers of lanes within each intersection are determined with an optimization model such that the traffic routing constraints are valid and certain demands and supplies are satisfied. The model of Step 3 is solved for every intersection $q \in \mathcal{J}$ of the considered street network with $\alpha_q \geq 2$.

Parameters and Sets

For every intersection $q \in \mathcal{J}$, $\alpha_q \geq 2$, the associated subgraph contains all intersection nodes $\mathcal{N}_{\mathcal{J}}(q)$, all arcs $\mathcal{A}_{\mathcal{J}}(q)$ of intersection q and all nodes and arcs directly connected to the intersection nodes in the detailed network $\mathbf{G} = (\mathcal{N}, \mathcal{A})$. Additionally a super source S_q, a super sink D_q and arcs connected with these super nodes are included depending on the flows of the commodities that result from Step 2.

Let $\mathcal{A}_C^{red}(co) := \mathcal{A}_C^{red} \cap \mathcal{A}_{used}^{red}(co)$ denotes the arcs of the street network for $co \in C_e$ and let $\mathcal{A}_C^{red}(veh) := \mathcal{A}_C^{red}$ denotes the arcs of the street network for the vehicles of evacuees in the reduced network $\mathbf{G}^{red}$. Note that the sets $\mathcal{S} := \bigcup_{co \in C} \mathcal{S}(co)$ and $\mathcal{D} := \bigcup_{co \in C} \mathcal{D}(co)$ of sources and sinks are defined in the reduced as well as in the detailed network. The components of the graph $\mathbf{G}_q = (\mathcal{N}_q, \mathcal{A}_q)$ of intersection q are defined as follows:

- $\mathcal{N}_{\mathcal{J}}(q) = \{1, \ldots, \alpha_q\}$ is the set of intersection nodes.
- $\mathcal{A}_{\mathcal{J}}(q) = \{(i, j) \in \{1, \ldots, \alpha_q\} \times \{1, \ldots, \alpha_q\} : i \neq j\}$ is the set of intersection arcs.
- Let the set $\mathcal{N}_P(q) = \{a \in \mathcal{S} \cup \mathcal{D} : q$ is adjacent to s in network $\mathbf{G}\}$ denotes the possible sources and sinks for intersection $q \in \mathcal{J}$.
- $\mathcal{A}_C(q)$ denotes the set of arcs that connect the nodes of the intersection q with nodes in $\mathcal{N}_P(q)$. They are contained in $\mathcal{A}_C$ of the detailed network ($\mathcal{A}_C(q) \subseteq \mathcal{A}_C$).
- $\mathcal{S}(q, co)$ denotes the set of the sources of commodity $co \in C$: it contains the nodes $s \in \mathcal{N}_P(q)$ with $\sum_{t \in T} x^{2nd}(co, s, q, t) > 0$ where $(s, q) \in \mathcal{A}_C^{red}(co)$.
- $\mathcal{D}(q, co)$ denotes the set of the sinks of commodity $co \in C$: it contains the nodes $d \in \mathcal{N}_P(q)$ with $\sum_{t \in T} x^{2nd}(co, q, d, t) > 0$ with $(q, d) \in \mathcal{A}_C^{red}(co)$.
- The set $\mathcal{A}_{S_q}(co) = \{S_q\} \times \mathcal{S}(q, co)$ denotes the set of arcs that leave super source S_q for $co \in C$.
- The set $\mathcal{A}_{D_q}(co) = \mathcal{D}(q, co) \times \{D_q\}$ denotes the set of arcs that lead to the super sink D_q for $co \in C$.

The optimization model of Step 3 is solved on the graphs $\mathbf{G}_q = (\mathcal{N}_q, \mathcal{A}_q)$ for every intersection $q \in \mathcal{J}$ ($\alpha \geq 2$) with the set of nodes $\mathcal{N}_q = \mathcal{N}_{\mathcal{J}}(q) \cup \mathcal{N}_P(q) \cup \{S_q, D_q\}$ and the set of arcs $\mathcal{A}_q = \mathcal{A}_{\mathcal{J}}(q) \cup \mathcal{A}_C(q) \cup \bigcup_{co \in C} (\mathcal{A}_{S_q}(co) \cup \mathcal{A}_{D_q}(co))$.

Step 2 provides for every intersection q the total flow of every commodity entering and leaving through the arcs in $\mathcal{A}_C^{red}$ that connect sources and sinks with intersection nodes. The results of Step 2 can be directly translated to the

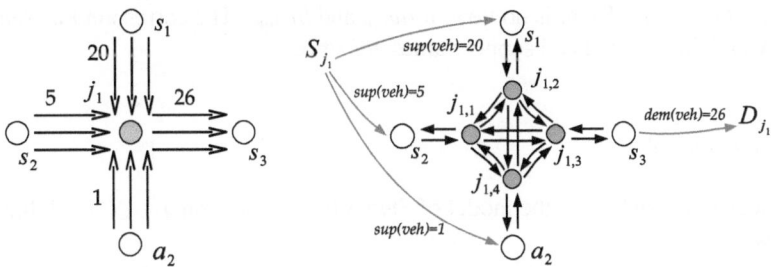

Fig. 8.5 Illustration of intersection j_1 of the 2×3-example in the reduced network $\mathbf{G}^{red}$ and in the network $\mathbf{G}_{j_1}$ in Step 3

entrances/exits of the intersection q leading to the amount of supply $sup(co, a)$ and demand $dem(co, a)$ for certain nodes $a \in \mathcal{N}_P(q)$ for commodity $co \in \mathcal{C}$. The additional demand and supply parameters of Step 3 (for a fixed intersection $q \in \mathcal{J}$) are defined as follows for $co \in \mathcal{C}$:

- $sup(co, a) := \sum_{t \in T} x^{2nd}(co, a, q, t)$ for the arc $(a, q) \in \mathcal{A}_{\mathcal{C}}^{red}(co)$ with node $a \in \mathcal{S}(q, co)$
- $dem(co, a) := \sum_{t \in T} x^{2nd}(co, q, a, t)$ for the arc $(q, a) \in \mathcal{A}_{\mathcal{C}}^{red}(co)$ with $a \in \mathcal{D}(q, co)$

Note, there is one arc $(a, q) \in \mathcal{A}_{\mathcal{C}}^{red}(co)$, $co \in \mathcal{C}$, and one arc $(q, a) \in \mathcal{A}_{\mathcal{C}}^{red}(co)$, $co \in \mathcal{C}$, for a fixed node $a \in \mathcal{N}_P(q)$ per construction of the representation of the street network.

Example of a Subgraph $\mathbf{G}_q = (\mathcal{N}_q, \mathcal{A}_q)$: $\mathbf{q} = \mathbf{j_1}$

Figure 8.5 illustrates such an subgraph of intersection j_1 of the 2×3-example. On the left, intersection j_1 is illustrated in the reduced network with the corresponding number of lanes and the total flow on the arcs of the commodity veh (results of Step 2). There are e.g. 20 vehicles that traverse arc (s_1, j_1) in the solution of Step 2 over the total time horizon and 26 vehicles that leave intersection j_1 using arc (j_1, s_3). The total flow of the commodity veh leading to and leaving the intersection in the optimal solution of the model of Step 2, indicated as sup and dem, are also depicted on the "super" arcs $\mathcal{A}_{S_q(veh)}$ and $\mathcal{A}_{D_q(veh)}$ on the graph $\mathbf{G}_{j_1}$ on the right. The nodes $j_{1,1}$, $j_{1,2}$ and $j_{1,4}$ are entrances for commodity veh and $j_{1,3}$ is an exit for commodity veh. In the solution of Step 3, 20 vehicles have to traverse arc (S_{j_1}, s_1) and 26 vehicles have to traverse arc (s_3, D_{j_1}), for example. The set of potential sources and sinks of the illustrated intersection j_1 is $\mathcal{N}_P(j_1) = \{s_1, s_2, s_3, a_1\}$ and the set of intersection nodes is $\mathcal{N}_{\mathcal{J}}(j_1) = \{j_{1,1}, j_{1,2}, j_{1,3}, j_{1,4}\}$. The set of sources of the vehicles in the subgraph $\mathbf{G}_{j_1}$ is $\mathcal{S}(j_1, veh) = \{s_1, s_2, a_1\}$ and the set of sinks is

$\mathcal{S}(j_1, veh) = \{s_3\}$. There is no flow of bus_{in} and bus_{out}. The corresponding sets are empty and the demand and supply values are zero.

Decision Variables

The decision variables of the model of Step 3 for intersection $q \in \mathcal{J}$ are defined as follows:

- $xJ(co, i, j, t)$ denotes inflow of commodity $co \in \mathcal{C}$ on arc $(i, j) \in \mathcal{A}_{\mathcal{J}}(q)$ at point in time t, where we set $xJ(co, i, j, t) = 0$ for all $co \in \mathcal{C}$, $(i, j) \in \mathcal{A}_{\mathcal{C}}(q)$: $(i \notin \mathcal{S}(q, co), j \in \mathcal{N}_{\mathcal{J}}(q))$ or $(i \in \mathcal{N}_{\mathcal{J}}(q), j \notin \mathcal{D}(q, co))$, $t \in \mathcal{T}$, and $xJ(co, i, j, t) = 0$ for all $t < 0$, $co \in \mathcal{C}$, $(i, j) \in \mathcal{A}_q$.
- $\ell_g(q)(i, j)$ denotes the number of lanes of type $g \in \{p, e\}$ within intersection $q \in \mathcal{J}$ for $(i, j) \in \mathcal{A}_{\mathcal{J}}(q)$.
- $yJ_g(i, j) \in \{0, 1\}$ denotes weather a positive number of lanes of type $g \in \{p, e\}$ is allowed $(= 1)$ to be assigned to arc $(i, j) \in \mathcal{A}_{\mathcal{J}}(q)$ or not $(= 0)$. For right turn arcs $(i, j) \in \mathcal{A}^r_{\mathcal{J}}(q)$, $y_e(i, j)$ is set to 0, because right turn arcs do not lead to crossing conflicts.
- $n_e(q) \geq 0$ denotes the number of emergency-lanes within the considered intersection q.

Objective Function

The objective is to minimize the weighted sum of flows leading to the super sink D_q incorporating the priority weight for the commodities and the number of emergency-lanes assigned in intersection q:

$$\min \sum_{co \in \mathcal{C}} \sum_{t \in \mathcal{T}} \sum_{d \in \mathcal{D}(co)} Pr_{co} \cdot t \cdot xJ(co, d, D_q, t) + P_q \cdot n_e(q) \qquad (8.11)$$

Flow Constraints

The flow conservation constraints and the constraints that avoid irregular flow within intersections and U-turns are analogously formulated to the corresponding constraints of the multicommodity evacuation model, i.e. the following constraints: the flow conservation constraints (7.2), "Avoidance of irregular behavior I" (7.3) and "Avoidance of irregular behavior II" (7.4).

- The flow conservation constraints have to be valid for every commodity $co \in \mathcal{C}$ for every node $i \in \mathcal{N}_{\mathcal{J}}(q) \cup \mathcal{N}_P(q)$ in every point in time $t \in \mathcal{T}$ (analogously to (7.2)).

- Avoidance of irregular behavior I (within intersection q) can be formulated as follows:

$$\sum_{(k,i)\in\mathcal{A}_{\mathcal{J}}(q)} xJ(co,k,i,t-\tau(co,k,i)) = xJ(co,i,j,t) \qquad (8.12)$$

for all $co \in \mathcal{C}$; $(i,j) \in \mathcal{A}_C(q) : i \in \mathcal{D}(q,co); t \in \mathcal{T}$

- Avoidance of irregular behavior II (along arcs $\mathcal{A}_C(q)$) can be formulated as follows:

$$xJ(co,i,j,t-\tau(co,i,j)) = \sum_{(j,k)\in\mathcal{A}_{\mathcal{J}}(q)} xJ(co,j,k,t) \qquad (8.13)$$

for all $co \in \mathcal{C}$; $(i,j) \in \mathcal{A}_C(q) : i \in \mathcal{S}(q,co); t \in \mathcal{T}$

- A connection between Steps 2 and 3: The flow determined with Step 3 should not travel faster through the intersection than in Step 2. Therefore the flow along the arcs $\mathcal{A}_C^{red}$ resulting from the optimal solution of Step 2 is taken into account.

$$\sum_{\tilde{t}=0}^{t} xJ(co,i,j,\tilde{t}) \leq \sum_{\tilde{t}=0}^{t} x^{2nd}(co,i^{red},j^{red},\tilde{t}) \qquad (8.14)$$

for all $co \in \mathcal{C}$; $(i,j) \in \mathcal{A}_C(q); t \in \mathcal{T}$

where $(i^{red},j^{red}) \in \mathcal{A}_C^{red}$ denotes the counterpart in the reduced network to arc $(i,j) \in \mathcal{A}_C(q)$ in the network $\mathbf{G}_q$ that models intersection q in detail and where we set $x^{2nd}(co,i^{red},j^{red},t) := 0$ for all $co \in \mathcal{C}$, $(i^{red},j^{red}) \in \mathcal{A}_C^{red} \setminus \mathcal{A}_C^{red}(co)$, $t \in \mathcal{T}$ for the ease of notation.

- The supplies and demands resulting from Step 2 have to be satisfied. The supply determined in Step 2 has to be dispatched: $sup(co,s)$ flow units of commodity $co \in \mathcal{C}$ have to start from the associated sources $s \in \mathcal{S}(q,co)$ within the time horizon $\mathcal{T}$

$$\sum_{t\in\mathcal{T}} xJ(co,S_q,s,t) = sup(co,s) \text{ for all } co \in \mathcal{C}; s \in \mathcal{S}(q,co) \qquad (8.15)$$

The demand determined in Step 2 has to be covered: $dem(co,d)$ flow units of commodity $co \in \mathcal{C}$ have to enter the sink $d \in \mathcal{D}(q,co)$ within the time horizon $\mathcal{T}$

$$\sum_{t\in\mathcal{T}} xJ(co,d,D_q,t) = dem(co,d) \text{ for all } co \in \mathcal{C}; d \in \mathcal{D}(q,co) \qquad (8.16)$$

Traffic Routing Constraints

The traffic routing has to be determined within the intersection. Outside the intersection, i.e. for the arcs in $\mathcal{A}_C(q)$, the number of lanes is already determined in Step 2 and it is fixed in Step 3. The constraints of Step 3 concerning the traffic routing are formulated with the decision variables $\ell_p(q)(i,j)$ and $\ell_e(q)(i,j)$ as well as the auxiliary variables $yJ_p(i,j)$ and $yJ_e(i,j)$ for all arcs $(i,j) \in \mathcal{A}_\mathcal{J}(q)$ within intersection q.

- The number of emergency- and public-lanes must not exceed the available number of lanes $\mathcal{L}(i,j)$, $(i,j) \in \mathcal{A}_\mathcal{J}(q)$ (analogously to constraint (7.14)).
- Crossing conflicts of public-lanes are prohibited and the number $n_e(q)$ of emergency-lanes assigned in intersection q is counted (analogously to constraints (7.15)–(7.17)).
- The number of mergings is restricted within intersection q (analogously to (7.18))

$$\sum_{(j,i) \in A_{\delta,q}^*} (\ell_e(q)(j,i) + \ell_p(q)(j,i)) \leq \ell^{2nd}(i^{red}, k^{red}) + (\delta - 1) \qquad (8.17)$$

$$\text{for all } i \in \mathcal{N}_\mathcal{J}(q); \ (i,k) \in \mathcal{A}_C(q); \ \delta \in \{1, \ldots, (\alpha_q - 1)\};$$

$$A_{\delta,q}^* \in \{A \subseteq \mathcal{A}_\mathcal{J}(q) : |A| = \delta, A \ni (j,i) \text{ for a node } j \in \mathcal{N}_\mathcal{J}(q)\}$$

where $(i^{red}, k^{red}) \in \mathcal{A}_C^{red}$ is the counterpart (in the reduced network $\mathbf{G}^{red}$) of arc $(i,k) \in \mathcal{A}_C(q)$ (in the network built in Step 3 for intersection q). The number of lanes $\ell^{2nd}(i^{red}, k^{red})$, $(i^{red}, k^{red}) \in \mathcal{A}_C^{red}$ are computed in Step 2 and they are parameters in Step 3.

Inflow and Total Capacity Constraints

The inflow and total street capacity constraints are formulated with inflow variables $xJ(co,i,j,t)$ for $(i,j) \in \mathcal{A}_\mathcal{J}(q) \cup \mathcal{A}_C(q)$, $co \in \mathcal{C}$, $t \in \mathcal{T}$, as well as the variables of number of emergency-lanes $\ell_e(q)(i,j)$ and public-lanes $\ell_p(q)(i,j)$ for $(i,j) \in \mathcal{A}_\mathcal{J}(q)$. For arcs $(i,j) \in \mathcal{A}_\mathcal{J}(q)$ within intersection q the inflow and total street capacity have to be determined with the number of lanes, i.e. the cumulated inflow of all commodities is bounded from above by $c^{in}(i,j) \cdot (\ell_e(q)(i,j) + \ell_p(q)(i,j))$ for every point in time and the inflow for vehicles within intersections is bounded by $c^{in}(i,j) \cdot \ell_p(q)(i,j)$ for every point in time (see inflow capacity constraints (7.19) and (7.20)). The total street capacity is determined analogously (see total street capacity constraints (7.22) and (7.23)).

For arcs in $\mathcal{A}_C(q)$ the capacities are determined with the number of lanes calculated in Step 2, i.e. the inflow capacity is $c^{in}(i,j) \cdot \ell^{2nd}(i^{red}, j^{red})$ and the total street capacity is $c(i,j) \cdot \ell^{2nd}(i^{red}, j^{red})$, where the arc $(i^{red}, j^{red}) \in \mathcal{A}_C^{red}$

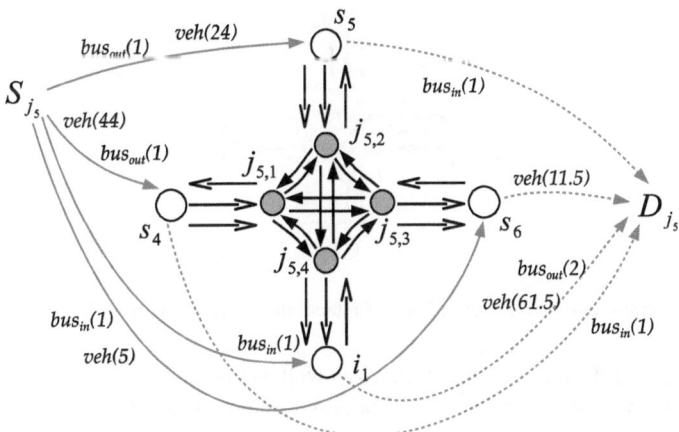

Fig. 8.6 Illustration of the graph $\mathbf{G}_{j_5}$ of j_5 in Step 3 with the corresponding demand and supply values

(in the reduced network that is considered in Step 2) is the counterpart of arc $(i, j) \in \mathcal{A}_C(q)$ (in the detailed network of intersection q that is considered in Step 3). The capacity constraints of the inflow and total street flow for arcs $(i, j) \in \mathcal{A}_C(q)$ are formulated analogously to constraints (7.21) and (7.24).

For all capacity constraints the appropriate consumption values $\varsigma_J^{in}(co)$, $\varsigma^{in}(co)$, $\varsigma_J(co)$ and $\varsigma(co)$ $(co \in \mathcal{C})$ have to be taken into account.

The 2×3-Example: Step 3

In the 2×3-example, there are six intersections that have four entrances/exits. For each of these intersections $j_1, j_2, \ldots, j_6$, the model of Step 3 has to be solved.

Let us consider intersection j_5 of example 2×3 for an illustration. In Fig. 8.6, the graph $\mathbf{G}_{j_5}$ associated with intersection j_5 is depicted. The filled arrows represent the defined arcs of the graph $\mathbf{G}_{j_5}$, i.e. the arcs within the intersections, the arcs emerging from the super source S_{j_5} (gray arcs) and arcs leading to the super sink D_{j_5} (gray arcs with a dashed line). The arcs $\mathcal{A}_C(j_5)$ that connect the sources and sinks with the nodes of the intersection are not illustrated. Instead, the corresponding fixed number of lanes are depicted (non-filled arcs) which are determined in Step 2. The values along the arcs $\mathcal{A}_{S_q}(j_5)$ denote the supplies of the commodities veh, bus_{in} and bus_{out} associated with the sources the arc is pointing at. The values next to arcs in $\mathcal{A}_{D_q}(j_5)$ denote the corresponding demands of the sinks which the arcs leave. The demands and supplies of the nodes for the commodities are determined with the results of the optimal solution of the model of Step 2. Node s_4 is a source for commodities veh and bus_{out} as well as a sink for commodity bus_{in}: In a feasible solution of the model of Step 3, 44 vehicles of evacuees and one bus bus_{out} have to travel through

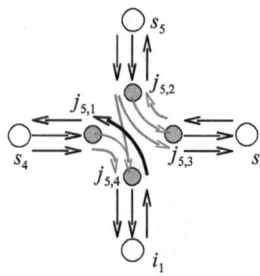

Fig. 8.7 Illustration of the number of lanes of intersection j_5 resulting from Step 3

the arc (S_{j_5}, s_4) during the time horizon as well as one bus bus_{in} has to traverse arc (s_4, D_{j_5}) during the considered time horizon (compare Fig. 8.6).

The number of lanes within intersection j_5 of the optimal solution of the Step 3 model are depicted in Fig. 8.7 where each arc represents a lane. One emergency lane is assigned to direction $(j_{5,4}, j_{5,1})$ of the intersection, it is a left turn. A crossing conflict occurs with the straight forward arc $(j_{5,2}, j_{5,4})$; it is $\ell_p(j_5)(j_{5,2}, j_{5,4}) = 1$.

8.4   Fourth Stage of the Multicommodity Heuristic

In Step 4 the resulting number of lanes generated with Steps 1, 2 and 3 are fixed in the urban multicommodity presented in Sect. 7.2. The inflow variables of all commodities are defined on the total detailed network **G** and have to be determined. The objective function (7.1) has to be minimized subject to the following constraints: the network flow constraints (7.2)–(7.4), the flow enforcements constraints concerning the vehicles of evacuees (7.5) and (7.6), the constraints concerning the buses (7.7)–(7.12) (including the constraints that provide possible starting points in time and that determine the number of buses that have to leave the assembly points) as well as the inflow and total street capacity constraints (7.19)–(7.24). The number of lanes that generate the traffic routing are fixed and are determined in Steps 1, 2 and 3. Feasibility of the values of the number of lanes is guaranteed, i.e. the traffic routing constraints (7.13)–(7.18) are valid.

With the fixed number of lanes, the second term of the objective function, the sum of penalty costs of assigning emergency-lanes, is already determined and fixed. In the heuristic approach, the emergency-lanes are not optimized considering the overall network, but only for every intersection separately.

The **2 × 3**-Example: Step 4

The optimal solution value of the model of Step 4 is 5, 233.17. The number of lanes in the optimal solution of example 2 × 3 are illustrated in Fig. 8.8. In total there are four emergency-lanes that are assigned in intersections j_2, j_5 and j_6. The evacuation

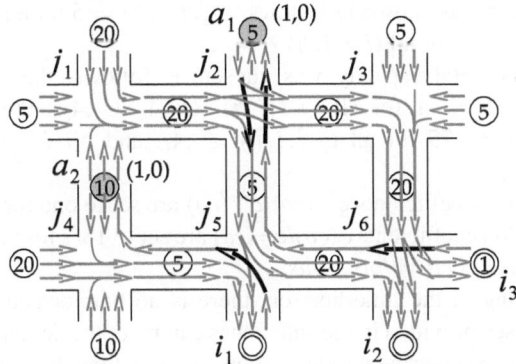

Fig. 8.8 Illustration of the number of lanes of the solution resulting from Steps 1, 2 and 3 of the 2 × 3-example

time is 107. The last vehicles arrive in the safe zones i_1, i_2 and i_3 at point in times 49, 50 and 45, respectively. The bus arrives in safe zone i_1 at point in time 107.

The bus enters the evacuation zone through safe zone i_3 and travels using intersections j_6, j_5, j_2 to assembly point a_1 to pick up evacuees. The evacuees are transported using intersections j_2 and j_5 to the safe zone i_1. The bus starts in i_1 again, traverses through intersections j_5 and j_4 to pick up the evacuees in assembly point a_2 to carry them back to safe place i_1.

8.5 Computational Study

The heuristic approach and the considered models are implemented with AMPL using gurobi 3.1. A computer running Windows XP with an AMD Athlon(tm) 64X2 Dual Core Processor 4600+ (2.41 GHz), 1.96 GB RAM was used for the computational study on the multicommodity urban evacuation model.

The computational study testing the four-staged multicommodity heuristic approach uses 150 instances that are based on small grid networks with intersections with $\alpha = 4$ entrances/exits. The sinks, the number of evacuees and the number of lanes are varied. The parameters of the instances were chosen as follows:

- Two different network sizes: a 2 × 3 grid network with one exit-pattern and three safe zones as well as a 4 × 5 grid network with four different exit-patterns ($k = 1, 2, 3, 4$) and five safe zones are considered. There are two fixed assembly points within the evacuation area in both networks. The sources and sinks of the buses in a safe zone are subsets of the provided safe zones.
- Ten patterns of number of evacuees randomly chosen from the set $\{5, 10, 15, 20\}$ are considered for instances based on the network 4 × 5 and the network 2 × 3.
- The number of usable lanes for all street sections was set to $\mathcal{L} = 2, 3$ or 4.

- The travel time τ was chosen as follows: $\tau(j,l,i,h) = 5$ for all $(j,l,i,h) \in \mathcal{A}_C$ and $\tau(i,l,i,h) = 1$ for all $(i,l,i,h) \in \mathcal{A}_{\mathcal{J}}$.
- The inflow and total capacity was chosen as follows: the inflow capacities are $c^{in}(j,l,i,h) = 1$ for all $(j,l,i,h) \in \mathcal{A}_C \cup \mathcal{A}_{\mathcal{J}}$, the total capacities are $c(j,l,i,h) = 75$ for all $(j,l,i,h) \in \mathcal{A}_C$ and $c(i,l,i,h) = 15$ for all $(i,l,i,h) \in \mathcal{A}_{\mathcal{J}}$.
- The consumption coefficients $\varsigma^{in}(co)/\varsigma_{\mathcal{J}}^{in}(co)$ are set to one for all $co \in \mathcal{C}$.
- The consumption coefficients $\varsigma(co)/\varsigma_{\mathcal{J}}(co)$ are set to 18/7 for vehicles $co = veh$ and 26/15 for buses $co = bus_{in}, bus_{out}$.
- At the beginning of the time horizon, there is no bus located at the assembly points. The buses provided in the safe zones can meet the demand of buses at the assembly points by entering the evacuation zone one time. Except in the instances based on the 2×3 network: There a provided bus has to enter the evacuation zone twice.
- The considered time periods are $\mathcal{T} = \{0, 1, \dots, 200\}$.
- The schedule is set to $\mathcal{BS}(s) = \{0, 5, 10, 15, \dots, 200\}$ for every source $s \in \mathcal{S}(co)$ of commodity $co \in \mathcal{C}_e$.
- The penalty costs and priority weights are set to one.

For comparison an LP-based relaxation is considered and the percentage gap

$$\% - \text{GAP} = \frac{UB - LB}{LB} \cdot 100$$

is considered, where UB denotes the weighted sum of flows that reach the corresponding sinks achieved with the four-staged heuristic approach and LB denotes the objective value of the LP-based relaxation of the multicommodity urban evacuation model. The weighted sum of flows that reach sinks is a term of the objective function, i.e. here the penalty for assigning emergency-lanes are disregarded. It is fixed in Step 4 with the fixed number of lanes resulting from Steps 1, 2 and 3.

The following LP-based relaxation of the urban multicommodity problem is considered to find fast a lower bound:

- The objective function that has to be minimized

$$\sum_{co \in \mathcal{C}} \sum_{d \in \mathcal{D}(co)} \sum_{t \in \mathcal{T}} Pr_{co} \cdot t \cdot x(co, d, D, t)$$

- Flow conservation constraints for all commodities (7.2)
- Avoidance of irregular flow I and II (7.3), (7.4)
- All vehicles have to leave the corresponding sources and have to leave the evacuation zone within the time horizon (7.5), (7.6)
- A possible bus schedule is provided (7.7), (7.8)
- The number of buses that have to leave assembly points, enter assembly points and leave the evacuation zone are modeled and restricted with constraints (7.9)–(7.12)

Table 8.1 Average percentage gaps between the results of the four-staged heuristic approach and the LP-based relaxation

Size	Exit-pattern k	Avg. %-GAP
2×3	1	11.92
4×5	1	19.05
4×5	2	10.67
4×5	3	26.08
4×5	4	11.09

Table 8.2 Average run-time of the computation of a solution of the four-staged heuristic approach and the computation of the LP-based relaxation in CPU-minutes

Size	Exit-pattern k	Avg. run-time four-staged appr.	Avg. run-time relaxation
2×3	1	5.04	11.26
4×5	1	153.35	38.80
4×5	2	128.11	24.42
4×5	3	119.04	30.07
4×5	4	103.02	7.05

- For the inflow capacity different cases depending on the commodity and on the arcs are taken into account. The flows on both directions are considered and they are restricted on the inflow capacity depending on the available number of lanes. Constraints (7.19) and (7.20) are changed to the following formulation:

$$\sum_{co \in \mathcal{C}} s_{\mathcal{J}}^{in}(co) \cdot (x(co, i, j, t) + x(co, j, i, t)) \leq c^{in}(i, j)\mathcal{L}(i, j) \qquad (8.18)$$

$$\text{for all } (i, j) \in \mathcal{A}_{\mathcal{J}}; t \in \mathcal{T}$$

Analog, the inflow capacity constraints for the arcs $\mathcal{A}_C$ are reformulated (7.21).
- The total street capacity takes both directions of a street section into account and restricts the flow on the total street capacity of the available number of lanes. I.e. the total capacity constraints (7.22)–(7.24) are reformulated analogously to the inflow capacity constraints.
- The decision variables of the relaxation are the inflow variables of all commodities, i.e. $x(co, i, j, t) \in \mathbb{R}_0^+$ for all commodities $co \in \mathcal{C}$ and for all arcs $(i, j) \in \mathcal{A}$ of the detailed network at the considered points in time $t \in \mathcal{T}$.

The results of the computational study are depicted in Tables 8.1–8.3. The average percentage gaps are aggregated over a fixed network size and a fixed exit-pattern. They are shown in Table 8.1.

The largest percentage gap occurs for an instance based on network 4×5 with $k = 3$ and $\mathcal{L} = 2$; the percentage gap is 49.44%. Such large percentage gaps occur for instances that have a relatively high number of safe zones where buses are located at the point in time $t = 0$ or for instances where multiple buses are assigned

Table 8.3 Average run-time of every step of the four-staged heuristic approach in CPU-minutes

Size	Exit-pat. k	Avg. run-time of Step 1	Avg. run-time of Step 2	Avg. run-time of Step 3	Avg. run-time of Step 4
2×3	1	0.06	3.68	1.30	0.84
4×5	1	0.61	123.66	6.87	22.21
4×5	2	0.52	120.18	7.41	18.12
4×5	3	0.99	112.51	5.54	10.38
4×5	4	0.48	97.55	4.99	5.94

to safe zones. The buses bus_{in} may take different routes to a sink (an assembly point within the evacuation area) in a solution of the Step 1 model. Along these routes that are directed in opposite direction of the evacuation flow, a lane is reserved in Step 2. That means these lanes are generally not used by evacuation flow and therefore between the solution of the four-staged heuristic approach and the solution of the relaxation may be a huge difference. In the instances based on network 4×5 with exit-patterns $k = 1$ and $k = 3$, the number of safe zones with a positive number of assigned buses bus_{in} (at point in time $t = 0$) are four. In the instances based on network 4×5 with exit-patterns $k = 2$ and $k = 4$, the safe zones where buses are located at the beginning of the time horizon are two and in instances based on the 2×3 grid network, there is only one safe zone where a single bus is located at point in time $t = 0$.

Considering the percentage gaps of the instances the following can be observed: 46.26% of the percentage gaps of the instances are less than or equal to 10%, 76.87% have percentage gaps that are less than or equal to 20% and 25.17% of the percentage gaps are greater than 20%.

The average computation time presented in Table 8.2 are aggregated over a fixed network size and a fixed exit-pattern. (The average computation times are given in "CPU-minutes", i.e. we consider CPU-seconds/60.) The number of buses that have to leave the assembly points and that are located at the beginning of the time horizon in safe zone are not varied for a network with a fixed exit-pattern. The time needed to run the four-staged heuristic approach is between 4 min and 247 min for a problem instance.

The average computation times for each of the four steps are aggregated over a fixed network size and a fixed exit-pattern. They are presented in Table 8.3. The time needed to run Step 1 on the considered instances lies between 3 to 60 s. Step 2 is the most time consuming step. The time to compute an instance of the model of Step 2 lies between 3 min and 238 min. In Step 3 an instance of every intersection has to be solved: in total the computation time of all models of Step 3 lies between 1 min and ca. 14 min and on average for every intersection around 0.29 min, i.e. 20 s. The computation time for the model in Step 4 with the fixed number of lanes lies between 0.15 min (i.e. 8 s) and 55 min.

Chapter 9
Conclusions and Future Research

This work provides different urban evacuation models that are based on dynamic network flow models. The decision variables are the flow of the evacuees and the traffic routing. The dynamic network flow models are extended with complicating constraints and integer variables and the flow is restricted depending on the assigned of number of lanes. Hence, the capacities of the network are not known in advance, but depend on the number of lanes that determine the traffic routing for the case of an evacuation. The underlying network maps the street network in detail. Every access of an intersection is modeled by a node and every turning direction is included in the set of arcs.

We focus on evacuation models assuming that the residents in the evacuation zone have the possibility to self evacuate. In Chap. 3, variations of a mixed-integer urban evacuation model are presented: different objectives and different possibilities to assign number of lanes (one-way/two-way) are considered as well as evacuation models with the capability of waiting on arcs and without waiting are provided.

The traffic routing for an evacuation is reorganized such that no crossing conflict of lanes occurs. Crossing conflicts are modeled dependent on turning directions, i.e. a direction-based formulation is chosen. With the observance of the constraints associated with the traffic routing and the assumption that every vehicle orders in the appropriate lane before entering an intersection, the lanes can be allocated such that no crossing conflict occurs. The introduced notation that uses two indices for a node provides the advantage that the crossing conflicts within intersections can be determined generally. The first index of an intersection node indicates a certain intersection and the second index denotes a certain entrance/exit of this intersection. With the second index, general formulations of constraints are possible to make the determination of the considered crossing conflicts comprehensible and reproducible. A contribution is that crossing conflicts for intersections with an arbitrary number of entrances/exits and arbitrary number of lanes can be identified with these constraints. The notation could be extended and used for a lane-based formulation instead of the direction-based one: three indices may be used for a node of an intersection that represents a connecting point of a certain lane to this intersection.

S. Bretschneider, *Mathematical Models for Evacuation Planning in Urban Areas*,
Lecture Notes in Economics and Mathematical Systems 659,
DOI 10.1007/978-3-642-28759-6_9, © Springer-Verlag Berlin Heidelberg 2013

These indices may denote the intersection, the entrance/exit of an intersection and the third index may indicate a certain lane adjacent to the entrance/exit of the considered intersection. However, the complexity of the approach would increase significantly.

In Chap. 5, the urban evacuation model is reformulated as a pattern-based model. The patterns can be translated as the union of number of lanes of certain parts of the network; street sections within an intersection (all turning directions) and street sections between two intersections. The patterns are considered for the one-way case and an approach for the generation of non-dominated and feasible patterns for this case is provided. Combinatorial aspects of crossing conflicts are considered where all crossing conflicts for arbitrary intersections for the one-way case are counted with respect to turning directions.

The mixed integer model is extended for an evacuation problem with multiple commodities. In particular, buses that pick up evacuees and carry them to safe zones are considered. These buses can drive multiple times from safe zones to assembly areas and back. New problems arise with this additional transportation mode and with the need to allow flow that travels in the opposite direction of the evacuation flow. The crossing conflict constraints have to be adjusted such that the feasibility for all reasonable instances can be guaranteed. A new type of lane is introduced within intersections: the emergency-lane that only emergency vehicles like buses or, if considered, ambulances are allowed to traverse. In some cases, if contraflow occurs, a crossing conflict has to be allowed such that a feasible solution in terms of the traffic routing can be generated.

The presented evacuation models are solved heuristically. For the different models heuristic approaches are developed, implemented and evaluated with computational studies. In Chap. 4, the relaxation-based approach is presented that solves first a relaxation of the mixed-integer urban evacuation problem and afterwards it repairs the infeasibilities for every intersection and neighboring arcs in terms of the constraints of the number of lanes. The pattern-based model is solved with a heuristic (Chap. 6) that takes different aggregation levels of the street network into account. First, the reduced street network that represents an intersection with a single node is considered and hence the constraints and decision variables that concern the turning directions within intersections are not needed. The results of the first step determine whether the intersection nodes are either entrances or exits. On that account the number of intersection patterns that are reasonable decreases. The pattern-based model is solved where only the reduced set of intersection patterns is considered and the traffic routing between intersections is fixed. A similar approach is used for the heuristic that solves the multicommodity model (Chap. 8). In different steps, different representations of the street network are taken into account: the reduced network, the intersection networks (a network per intersection) and the detailed network. The model is solved on the reduced network disregarding the intersection constraints and the corresponding decision variables where first only the flows of buses (Step 1) and afterward the flows of all commodities (Step 2) are considered. In a third step, when using the results of the reduced network for the detailed network, for every intersection a network is generated and for each of them an optimization model is solved separately. The complexity is reduced compared to the complexity if considering all intersections in the same time, i.e. the entire detailed network. On the other hand, dependencies between intersections cannot be incorporated in the same way. In Steps 1 through 3 a feasible traffic routing is generated and in Step 4, fixing the achieved traffic routing, the flows of evacuees and buses as well as the objective function value are computed.

In addition to the traffic routing, a result of the solution of the models is an estimation of the evacuation time. The estimated evacuation time can be used in deciding whether an evacuation is adequate for a certain situation or if another measure like in-place sheltering may be a better alternative. In case of a chemical accident, the estimated evacuation time is an important information besides, e.g., the information what kind of chemical accident occurred or the current and predicted wind direction. It supports the decision whether to order an evacuation or recommend that people stay in houses and seal windows and doors appropriately.

Practical questions arise with the implementation of the solution of the models: There is always the problem or question of compliance or non-compliance of the evacuees if providing a new traffic routing. Trained personnel are needed to enforce and implement the changed traffic routing. The population needs to be informed about the measures that are taken in case of an evacuation and also about the corresponding changes in the traffic routing. Simulation approaches that take the aspect of compliance into account may be used to evaluate these results (see e.g. Pel et al., 2010). Overall, the size of the time-expanded networks growths especially when the time horizon has to be high (the complete network has to be cleared). This may lead to practical computational problems like the lack of memory or inappropriate long computation times (those problems arose in the computational studies, see e.g. Sect. 6.3). The considered street network may be only a subnetwork of the existing one; only arterials and main street may be considered such that the organizational effort as well as the complexity of the problem are restricted and do not become unnecessary large. The traffic routing of the street network of the side roads could be almost the same as usual. Accesses and adjacent road ways to the traffic routing that is changed for the evacuation may be varied and/or restricted as well.

Multiple variations and extensions of the proposed evacuation models and heuristic solution approaches are possible. These variations and extensions could be compared to the presented approaches in terms of the computational times and the results like the value of the objective functions or the values of the decision variables.

The two-staged heuristic approach proposed for the pattern-based model could be adapted to the one-way mixed-integer formulation of the urban evacuation model. That means, the model of Step 1 of the two-staged approach uses a mixed-integer formulation based on the reduced network and in Step 2 the results of Step 1 are used to restrict the turning arcs within intersections. Flow is only defined on arcs that lead from an entrance node to an exit node of an intersection and therefore the possibilities to assign number of lanes are reduced. An interesting computational study would be to compare the two-staged approaches of the mixed-integer formulation and the pattern-based formulation in run-time and results.

In some of the problem instances, the multicommodity heuristic approach generates results with relatively high percentage gaps between the constructed feasible solution and the considered lower bound. We suggest varying the problem of the first step of the multicommodity approach. The first step determines the routes of the buses without considering the vehicles of evacuees and pass lower bounds for the number of lanes to the second step. In some instances "too many" different routes lead from safe zones to assembly areas, i.e. the assigned lanes are directed in opposite direction to the evacuation flow. Therefore "too many" lanes (depending on the considered problem instance) are reserved for directions that are opposite to the evacuation flow and may lead to a feasible solution with a relatively high percentage gap. To avoid such results, the total number of lanes could be restricted in the first step. For example, in the objective function, costs of number of lanes could be taken into account. Also more restrictive constraints are possible to implement: already in Step 1 the routes for buses could be reduced or the assembly areas could be restricted to which buses from certain safe places are permitted to drive. These constraints could be realized using another formulation for the network flow, a path flow formulation that considers only predefined routes.

That leads us directly to another aspect that could be exchanged in the studied optimization models: the arc flow formulation describing the traffic flow. Instead, different approaches that describe the network flow could be used. There are, for example, the above mentioned path flow formulation or formulations that model flow-dependent transit times (see Sect. 2.1). An interesting question to study may be if and how different network flow formulations influence the resulting traffic routing, i.e. whether a network flow formulation that captures certain traffic effects better/differently than the dynamic network flow formulation leads to different results in the traffic routing.

Moreover, the evacuation problems may be extended in several ways: More commodities than the considered ones can be included like trains, pedestrians or ambulances. Evacuation plans for nursing homes, hospitals or schools can be taken into account with the incorporation of routes of ambulances and buses that are needed to evacuate these locations. Capacities of destinations and the associated shelters may be considered. Not all safe zones have the same capacity of food, water or sleeping-places. Hence it may be desirable to lead different amounts of evacuees to different destinations. The traffic routing could be restructured such that also the aspect of the capacity of the sinks that lead to certain shelters is regarded. There are already approaches that model the capacity of sinks (see Sect. 2.3), but which do not consider a traffic routing with the capability of lane-reversal. Staged evacuation is used when the impact of a hazard hits different parts of the evacuation area with different levels of intensity at different points in time. Generally, parts that are affected earlier than the other ones are evacuated prior. Staged evacuation may be taken into account while determining the traffic routing: the area may be divided into parts of towns, districts, a couple of neighboring streets, city blocks or zipcode-wise and a sequence may be determined in which ordering these parts have to be evacuated. Staged evacuation is considered for a fixed traffic routing, for example, in Liu et al. (2006) (an optimization model) or Chen and Zhan (2008) (a simulation approach).

Additionally, effects on the traffic routing may be studied that result from stochastic events or uncertainties. These may be accidents, storm surges or heavy rain fall that could influence the inflow and total street capacity as well as the travel time. Or this may be uncertainty concerning other parameters like the number of evacuees. This is a parameter of the urban evacuation models assigned to each source. The uncertainties may be directly incorporated in the optimization model with tools of stochastic optimization (see e.g. Ng and Waller, 2009) or robust optimization (see e.g. Yao et al., 2009). Alternatively, the effects to the results of the evacuation models may be also evaluated with a suitable sensitivity analysis.

Generally, we would not include all of the above-mentioned issues in the same model because of complexity. However, it may be an interesting topic to study the effects of different additional aspects or variations of the traffic flow formulation to the results of the urban evacuation models.

References

Chen X, Zhan F (2008) Agent-based modelling and simulation of urban evacuation: relative effectiveness of simultaneous and staged evacuation strategies. J Oper Res Soc 59:25–33

Liu Y, Lai X, Chang G-L (2006) Cell-based network optimization model for staged evacuation planning under emergencies. Transport Res Rec: J Transport Res Board 1964:127–135

Ng M, Waller ST (2009) The evacuation optimal network design problem: model formulation and comparison. Transport Lett: Int J Transport Res 1:111–119

Pel AJ, Hoogendoorn SP, Bliemer MC (2010) Evacuation modeling including traveler information and compliance behavior. In: Hoogendoorn SP, Pel AJ, Taylor MAP, Mahmassani H (eds), Proceedings of the 1st International Conference on Evacuation Modeling and Management, Elsevier Procedia Engineering, vol 3, http://www.sciencedirect.com/science/article/pii/S1877705810004807, 20 Feb 2011, pp 101–111

Yao T, Mandala SR, Chung BD (2009) Evacuation transportation planning under uncertainty: a robust optimization approach. Network Spat Econ 9:171–189

Erratum to: Mathematical Models for Evacuation Planning in Urban Areas

Sarah Bretschneider

Erratum to:
S. Bretschneider, Mathematical Models for Evacuation Planning in Urban Areas,
Lecture Notes in Economics and Mathematical Systems 659, DOI 10.1007/978-3-
642-28759-6 © Springer-Verlag Berlin Heidelberg 2013

The publisher regrets the following errors occurred in the book *Mathematical Models for Evacuation Planning in Urban Areas by Sarah Bretschneider.*

The below Text was missing in the book. The text is printed here:

This work "Mathematical Models for Evacuation Planning in Urban Areas" is an approved dissertation for the candidature for the academic degree "Dr. rer. nat." by the Faculty of Mathematics of the University of Duisburg-Essen.
Names of the referees:
1. Prof. Dr. Rüdiger Schultz
2. Prof. Dr. Alf Kimms
3. Prof. Dr. Sigrid Knust
Date of the oral doctorate defense: 20th of September 2011.

Diese Arbeit "Mathematical Models for Evacuation Planning in Urban Areas" wurde von der Fakultät Mathematik der Universität Duisburg-Essen als Dissertation zur Erlangung des Doktorgrades (Dr. rer. Nat.) genehmigt.
Name der Gutachterinnen und Gutachter:
1. Prof. Dr. Rüdiger Schultz
2. Prof. Dr. Alf Kimms
3. Prof. Dr. Sigrid Knust
Tag der Disputation: 20. September 2011

The original online version for this book can be found at
http://dx.doi.org/10.1007/978-3-642-28759-6

Appendix A
Information About the Computational Study
of the Relaxation-Based Approach

A.1 Reference Values Used to Compare the Results of the Relaxation-Based Approach

In Table A.1 the reference values for the computational study in Chap. 4 are illustrated. Each reference value is indicated by "opt", "LB (mipgap value·100)", or "relax". The gap (illustrated within in the round brackets) is rounded to the second decimal place and is the relative gap that is an output from AMPL. In the second and third columns the results concerning the instances with exit-pattern $k = 1$ are depicted and in the fourth and fifth column the results concerning the instances with exit-pattern $k = 2$ are illustrated.

A.2 Network Representation

The representation of the network used in the computational study in Sect. 4.3 varies slightly from the representation presented in Sect. 3.1.

The arcs next to a source on the boundary were used to construct the network. A blind alley is interpreted as an intersection with $\alpha = 1$ entrances/exits. Between two intersections where non of them are declared as a destination a source is built. For sources on the boundary only the arcs leading from the source to the network are needed. The arcs that are adjacent to the intersection with $\alpha = 1$ (in Fig. A.1 the gray nodes on the boundary that indicate entrances and exits of intersection) are not needed. These arcs are never traversed by flow and are therefore no longer considered in the network representation in Sect. 3.1.

S. Bretschneider, *Mathematical Models for Evacuation Planning in Urban Areas*,
Lecture Notes in Economics and Mathematical Systems 659,
DOI 10.1007/978-3-642-28759-6, © Springer-Verlag Berlin Heidelberg 2013

Table A.1 Comparing the results of the heuristic and best known lower bound where T = 150 (++ indicates instances with T = 200)

Instance	%-GAP	Reference value	%-GAP	Reference value
Network	3 × 4_1		3 × 4_2	
La2 Pers10	0.49	opt	3.69	opt
La2 Pers30	4.66	LB (1.4)	0.28	opt
La2 PersMix	0.25	LB (0.01)	12.57	opt
La3 Pers10	0.06	opt	0.62	opt
La3 Pers30	5.16	LB (0.49)	6.87	opt
La3 PersMix	0.15	LB (0.20)	2.13	opt
La4 Pers10	0.06	LB (<0.01)	0.52	opt
La4 Pers30	12.75	LB (2.9)	0.57	opt
La4 PersMix	0.29	opt	0.06	opt
Network	4 × 5_1		4 × 5_2	
La2 Pers10	36.60	LB (2.6)	0.00	opt
La2 Pers40	0.40	relax	7.80	LB (0.96)
La2 PersMix	16.03	LB (9.27)	3.39	opt
La3 Pers10	8.47	LB (1.8)	0.00	opt
La3 Pers40	5.62	relax	4.34	LB (0.95)
La3 PersMix	6.28	LB (7.02)	1.22	opt
La4 Pers10	2.14	LB (1.19)	0.11	LB (0.01)
La4 Pers40	7.33	relax	13.39	relax
La4 PersMix	8.26	LB (4.21)	0.10	LB (<0.01)
Network	5 × 5_1		5 × 5_2	
La2 Pers10	0.00	relax	2.51	opt
La2 Pers30	0.00 ++	relax	2.35	relax
La2 PersMix	0.86	LB (9.44)	3.54	opt
La3 Pers10	5.94	LB (2.06)	0.11	opt
La3 Pers30	0.00	relax	5.51	LB (1.25)
La3 PersMix	11.97	relax	1.11	LB (<0.01)
La4 Pers10	0.84	LB (1.57)	0.04	opt
La4 Pers30	0.00	LB (2.44)	10.49	LB (0.77)
La4 PersMix	21.86	relax	2.40	LB (<0.01)
Network	5 × 6_1		5 × 6_2	
La2 Pers10	0.00	relax	0.24	LB (≪0.01)
La2 Pers30	0.00 ++	relax	3.39	relax
La2 PersMix	1.93	LB (6.87)	6.61	opt
La3 Pers10	1.35	LB (2.40)	0.00	LB (0.17)
La3 Pers30	0.00	relax	9.70	LB (1.52)
La3 PersMix	18.02	relax	0.22	LB (<0.01)
La4 Pers10	0.71	relax	0.21	relax
La4 Pers30	0.00	LB (1.10)	8.37	relax
La4 PersMix	10.37	relax	0.03	opt

(continued)

Table A.1 (continued)

Instance	% GAP	Reference value	%-GAP	Reference value
Network	$5 \times 7\_1$		$5 \times 7\_2$	
La2 Pers10	10.13	relax	0.81	relax
La2 Pers30	0.00 ++	relax	4.49	relax
La2 PersMix	10.17	relax	0.71	LB (0.05)
La3 Pers10	3.29	LB (2.67)	0.07	relax
La3 Pers30	0.00	relax	10.44	relax
La3 PersMix	7.96	relax	2.47	LB (<0.01)
La4 Pers10	0.67	relax	0.18	relax
La4 Pers30	0.00	relax	7.17	relax
La4 PersMix	39.72	relax	0.00	opt

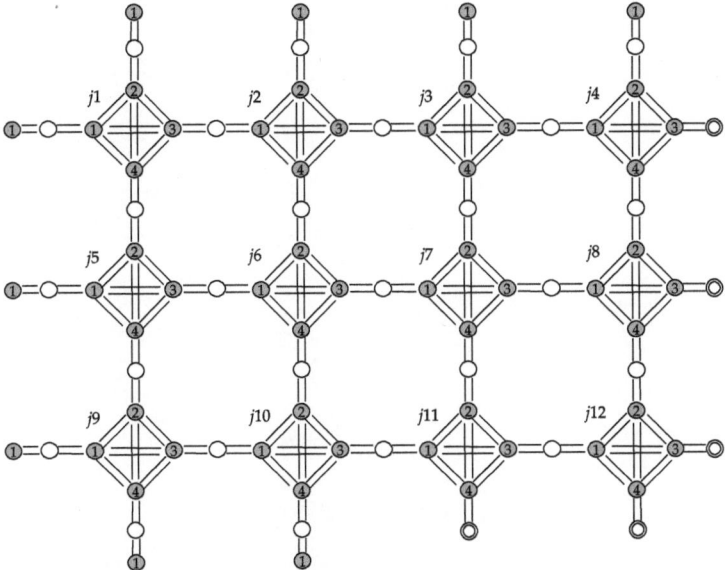

Fig. A.1 Grid network of size 3×4 with exit-pattern $k = 1$ with superfluous arcs

See for an illustration and comparison Fig. A.1. A 3×4 network is shown with five sinks and nine intersections with $\alpha = 1$ entrances/exits on the boundary of the network.